Comparative Behavior of African Monkeys

MONOGRAPHS IN PRIMATOLOGY

Volume 1: Child Abuse: The Nonhuman Primate Data
Martin Reite and Nancy G. Caine, *Editors*

Volume 2: Viral and Immunological Diseases in Nonhuman Primates
S.S. Kalter, *Editor*

Volume 3: Blood Groups of Primates: Theory, Practice, Evolutionary Meaning
Wladyslaw W. Socha and Jacques Ruffié

Volume 4: Female Primates: Studies by Women Primatologists
Meredith F. Small, *Editor*

Volume 5: Clinical Management of Infant Great Apes
Charles E. Graham and James A. Bowen, *Editors*

Volume 6: Nonhuman Primate Models for Human Growth and Development
Elizabeth S. Watts, *Editor*

Volume 7: The Lion-Tailed Macaque: Status and Conservation
Paul G. Heltne, *Editor*

Volume 8: Behavior and Pathology of Aging in Rhesus Monkeys
Roger T. Davis and Charles W. Leathers, *Editors*

Volume 9: Primate Conservation in Tropical Rain Forest
Clive W. Marsh and Russell A. Mittermeier, *Editors*

Volume 10: Comparative Behavior of African Monkeys
Evan L. Zucker, *Editor*

Comparative Behavior of African Monkeys

Editor

Evan L. Zucker
Department of Psychology
Loyola University
New Orleans, Louisiana

ALAN R. LISS, INC., NEW YORK

Address all Inquiries to the Publisher
Alan R. Liss, Inc., 41 East 11th Street, New York, NY 10003

Printed in the United States of America.

Library of Congress Cataloging-in-Publication Data

Comparative behavior of African monkeys.

(Monographs in primatology; v. 10)

Includes bibliographies and index.

1. Cercopithecus—Africa—Behavior. 2. Erythrocebus—Africa—Behavior. 3. Mammals—Behavior. 4. Mammals—Africa—Behavior. I. Zucker, Evan L. II. Series. [DNLM: 1. Behavior, Animal. 2. Cercopithecidae—physiology. 3. Cercopithecus—physiology. 4. Social Behavior. W1 M0568N v. 10 / QL 737.P93 C737]
QL737.P93C58 1987 599.8′2 87-16922
ISBN 0-8451-3409-4

Contents

Contributors

Mollie Bloomstrand, University of Texas System Cancer Center, Bastrop, TX 78602 **[197]**

Claud A. Bramblett, Department of Anthropology, The University of Texas at Austin, Austin, TX 78712 **[67]**

Margaret R. Clarke, Department of Comparative Medicine, Wake Forest Medical Research Center, Winston-Salem, NC 27103; present address: Delta Regional Primate Research Center of Tulane University, Covington, LA 70433 **[81]**

Anthony M. Coelho, Jr., Laboratory of Behavioral Medicine, Southwest Foundation for Biomedical Research, San Antonio, TX 78284 **[67]**

Carolyn M. Crockett, Department of Psychology, University of Washington, and Woodland Park Zoological Gardens, Seattle, WA 98195 **[39]**

J. Erwin, National Geographic Society, Washington, DC 20035 **[1]**

Anna R. Grady, Department of Comparative Medicine, Wake Forest Medical Research Center, Winston-Salem, NC 27103 **[81]**

Jay R. Kaplan, Department of Comparative Medicine, Bowman Gray School of Medicine of Wake Forest University, Winston-Salem, NC 27103 **[81,127]**

Barbara C. Kirkevold, Department of Psychology, University of Washington, and Woodland Park Zoological Gardens, Seattle, WA 98195 **[39]**

James Loy, Department of Sociology and Anthropology, University of Rhode Island, Kingston, RI 02881 **[23,175]**

Kent M. Loy, Heaton Orchard Road, West Kingston, RI 02892 **[23]**

Terry L. Maple, School of Psychology, Georgia Institute of Technology, Atlanta, GA 30332 **[197]**

Elizabeth C. Nicks, Department of Comparative Medicine, Wake Forest Medical Research Center, Winston-Salem, NC 27103 **[81]**

Evan L. Zucker, Department of Psychology, Loyola University, New Orleans, LA 70118 **[ix,1,99,151]**

Preface

The production and publication of the **Comparative Behavior of African Monkeys** were undertaken to fulfill two purposes. First, this volume contains data regarding the social behavior of several species of monkeys in the genera *Cercopithecus* and *Erythrocebus*. Secondly, and more importantly, the behavior of these monkeys is analyzed and interpreted comparatively. Valid generalizations regarding primate behavior depend upon data that are comparative and integrative. A comparative approach to the study of behavior can take several forms, and the contributors to this volume were encouraged to take whichever approach they felt was the most appropriate for their work. However, a comparative approach was mandatory. The comparative perspectives included in this volume are: Across species within the same genus, as in comparisons between two or more species of *Cercopithecus* monkeys; across genera to other African monkeys, such as in the comparisons of a *Cercopithecus* species with *Erythrocebus* or a *Papio* species; and across taxa to other species of Old World monkeys for whom comparative data are available, such as comparisons between a *Cercopithecus* species and a *Macaca* species. In addition, comparisons across living conditions are included, assessing similarities or differences between feral and captive African monkeys. The comparisons made by the contributors are indicated in Table 1.

Table 1. Primary Comparisons Made by Contributors

Chapter #	Author(s)	Within Genus	Intergeneric, African	Intergeneric, Asian	Captive-Feral
1	Erwin and Zucker	X	X	X	
2	Loy and Loy		X		X
3	Kirkevold and Crockett	X	X	X	X
4	Bramblett and Coelho	X	X		
5	Clarke et al		X		X
6	Zucker		X	X	
7	Kaplan		X	X	
8	Zucker		X	X	
9	Loy	X	X		
10	Bloomstrand and Maple	X	X		X

It was not intended for this volume to contain empirical data for every species of African monkey, nor to include comparative analyses that were totally comprehensive. The number of possible comparisons, given the guidelines delineated above, would be enormous. The comparisons that were to be included were those that would be the most appropriate and meaningful rather than simply numerous, and were to be made along lines of phylogeny, social systems, and types of behaviors. The chapters in this volume are a mixture of empirical findings and reviews, and are focused primarily on several species in the genus *Cercopithecus*, as well as *Erythrocebus*.

The starting point for this volume was a symposium organized for the IXth Congress of the International Primatological Society. Four of the papers that were presented in that symposium have been expanded, updated, and included in this volume. Six additional chapters have been included, all written since the original symposium, and updated during the revision process.

The trends of the first five years of the current decade, in terms of the African monkey species used in behavioral research, are addressed by Erwin and Zucker in the opening chapter. This first chapter was intended to summarize the recent past, as well as to highlight those species that are relatively unknown behaviorally, and to encourage additional behavioral research.

The next eight chapters that follow are arranged developmentally, beginning with three chapters describing the social and behavioral development of infants. Loy and Loy, in Chapter 2, describe the early social development of patas monkeys housed in an outdoor corral. Their focus is the development and expression of sex differences during the first year of life of infant patas monkeys. In Chapter 3, Kirkevold and Crockett describe the development of zoo-living DeBrazza's monkeys, particularly the infants' interactions with other group members and the attainment of independence. The behavioral development of captive vervets, Sykes' monkeys, and savanna baboons during their first seven years of life is addressed by Bramblett and Coelho in Chapter 4. Differences in the timing at which immatures begin to approximate adult patterns of behavior are highlighted, and are discussed with respect to the ecological factors that might contribute to these different rates of development.

Chapter 5, by Clarke, Kaplan, Nicks, and Grady, is a "transition" chapter developmentally. In this chapter, Clarke et al report their observations on how young patas and African green monkeys, living in a mixed-species corral, differ in their use of available space, and how these differences in "arboreality" and "terrestriality" influence the social interactions that occur between these two species. Although some of the subjects studied by Clarke et al were younger than those studied by Loy and Loy, Kirkevold and Crockett, and Bramblett and Coelho, these young patas and African green monkeys were living in a large peer group, independent of their mothers. Thus, behaviorally, they resemble juveniles more so than infants.

Chapters 6 through 9 focus primarily on the behavior of adult African monkeys. The interspecific nature of Clarke et al's chapter is continued by Zucker with his chapter detailing the polyspecific and interspecific interactions of free-ranging patas and rhesus monkeys that were maintained on the island of El Guayacan, at La Parguera, PR. Whereas these species are not sympatric naturally, they both needed to adapt to the problems associated with sharing resources with an alien species. The patterns of interactions observed are examined in terms of the ecological variables applied to naturally-occurring mixed-species interactions, as well as the behavioral and communicative factors that were operating.

In Chapter 7, Kaplan compares the parameters of several types of social interactions for several species of Old World monkeys, including patas, vervets, and rhesus. The affiliative behaviors of these species are viewed in terms of the expression of dominance by these species.

Zucker, in Chapter 8, investigates the distribution of social behaviors among females as a function of the rank disparities between the initiators and recipients. Female-female grooming and allomaternal behaviors are considered comparatively, as well as the occurrence and patterning of intergender agonism and grooming. The primary comparisons made involve patas, vervets, bonnets, and rhesus females.

In Chapter 9, Loy critiques the concept of estrus and its use as a predictor of sexual behavior for several African monkey species. Loy discusses the utility of the ARP model of sexual behavior (attractivity, receptivity, and proceptivity), and encourages the use of this framework in the analysis of sexual behavior.

Chapter 10 also deals with sexual behavior, but in this chapter, Bloomstrand and Maple discuss sexual behavior as it relates to the management and successful reproduction of African monkeys in captivity. They examine such reproductive parameters as cycle length, conception rates, breeding strategies, fetal wastage, and interbirth intervals. Most importantly, they consider these parameters in light of the social structures and behavior of the respective species, making suggestions and recommendations that are species-appropriate.

Examination of the contents of this volume reveals that the data included in these chapters were obtained from monkeys in several types of captive environments, including zoos, research centers, and breeding colonies, as well as from monkeys that were free-ranging. Although there are no chapters dealing *directly* with the behavior of monkeys living in feral groups, the extensive comparisons made by these authors have resulted in the inclusion of field data in nearly every chapter (if not in every chapter).

The comparative perspective that has guided the development of this volume follows from that of Hodos and Campbell (1969). Hodos and Campbell, among others, such as Erwin (1981), advocated that the species used in comparative research be selected either on the grounds of phylogenetic relatedness, or on the basis of adaptations that different species have evolved to meet the challenges of similar (or common) environments. This view of comparative research has continued to provide direction for my own work and professional development. As a freshman in Hodos' Introductory Psychology class in the spring of 1972 at the University of Maryland, I had no idea of what was to follow. One thing that did follow was that this comparative perspective became even more engrained several semesters later when I was a student in Hodos' Animal Behavior course.

In addition to this expression of thanks to William Hodos, I also wish to thank those that have contributed to the production of this volume, particularly Ann Epner, Joe Erwin, the production staff at Alan R. Liss, Inc., and Margaret Clarke. My student assistant, Rose McCleary, has been invaluable. I thank all of these people for their patience with me, and I thank the contributors, as well, for their continued patience. The day-to-day demands of academe, especially at universities devoted to undergraduate education, are not always conducive to projects such as this. This volume has taken a long time to compile, but I sincerely hope that readers find it worthwhile.

Evan L. Zucker

References

Erwin J (1981): The comparative nature of primate biology. Am J Primatol 1: 125–127.

Hodos W, Campbell CBG (1969): Scala Naturae: Why there is no theory in comparative psychology. Psychol Rev 76: 337–350.

Comparative Behavior of African Monkeys, pages 1–21

1

African Monkeys in Behavioral Research: A 5-Year Retrospective Analysis

J. Erwin and Evan L. Zucker

National Geographic Society, Washington, DC 20035 (J.E.), and Department of Psychology, Loyola University, New Orleans, LA 70118 (E.L.Z.)

INTRODUCTION

The focus of this volume is the social behavior of several species of African monkeys, investigated and interpreted with a comparative perspective. The development of knowledge of these African species has temporally paralleled the acquisition of knowledge of other primate species, and in order to place recent research trends in "temporal perspective," it is important to consider the historical development of the research on these African species of nonhuman primates, in both field and laboratory settings. Following this brief historical review, an analysis of the trends in species use is presented. This analysis covers the 5-year period from 1980 through 1984. Data are included for the genera *Cercocebus*, *Cercopithecus* (including *Miopithecus* and *Allenopithecus*), *Colobus*, *Erythrocebus*, and *Papio*, those species included under the heading of "African monkeys." Comparisons, by species, are included for the two genera that contain the most species, i.e., *Cercopithecus* and *Papio*. Although Barbary macaques are African, they are included with the other species of *Macaca*. The macaques (genus *Macaca*) and Asian leafeating monkeys of the genus *Presbytis* provide a comparative context for examination of patterns of research with African monkeys.

In establishing a historical perspective, it is difficult to identify with certainty a single study as the first of its kind. The historical information presented here addresses primarily the "modern era" [Maple, 1979, p. 48], but is not limited to this post-1940 period. For additional information on the history of primate studies, see Maple [1979], and Southwick and Smith [1986] for a history of field studies.

Field Research

Guenons and colobines. Williams [1981] compiled a bibliography containing citations of behavioral studies of feral and free-ranging guenons and vervets. According to Williams [1981], the earliest publication on the behavior of any of these species was by B. L. Mitchell in 1946; it dealt with vervets (*Cercopithecus aethiops*) and blue monkeys (*C. mitis*). Shortly thereafter, another paper appeared [Page, 1949], and several more were published in the early 1950s [i.e., Buxton, 1951, 1952; Lumsden, 1951; Haddow, 1952; Watson, 1951]. Reports, such as those by Buxton [1951] and Haddow [1952] were linked closely to epidemiological studies of yellow fever in Uganda. The initial ecological and behavioral work on *Colobus* has been attributed to A. H. Booth [1957] by Southwick and Smith [1986], and C. Booth published a report of her observations of three species of guenons 5 years later [Booth, 1962].

Using the entries in Williams [1981] to estimate the relative attention received by the various species, we found that approximately 39% of the entries involved either *C. aethiops* or *C. mitis*, the two species observed and reported on by Mitchell [1946]. These are the two species of guenons most extensively studied. Trends based on the year of publication were generated from Williams's [1981] bibliography. One third of the entries were works published prior to 1971 [1946–1970], with the remaining two thirds distributed equally over the two most recent 5-year blocks of time (1971–1975 and 1976–1980, inclusive). It is interesting to note that just over half (88) of the 168 entries in Williams's [1981] bibliography were publications that focused on only one guenon species; eleven additional citations were of studies of talapoins, and the remainder were studies of general African habitat usage, studies of African apes, reviews of topical areas, or textbooks. Thus, there was relatively little direct attention afforded these species of African monkeys; rather, they were studied as a part of a larger, ecologically-based context.

Patas monkeys. The pioneering field work on patas monkeys was begun by Hall in mid-1963 [Hall, 1965], and was continued at other sites by Struhsaker and Gartlan during 1966–1968 [Struhsaker and Gartlan, 1970]. As these monkeys are rather elusive relative to such species as baboons, behavioral data from wild patas remain relatively scarce (see analyses presented in this chapter). Hall established a small captive colony of patas in Bristol, and wrote several papers about these groups before his death in 1965

[Hall, 1967; Hall et al, 1965; Hall and Goswell, 1964; Hall and Mayer, 1966, 1967].

Baboons. Long-term field studies of several species of baboons began in the late 1950s in several different locations. Bolwig [1959] began his observations of captive and wild chacma baboons (*Papio ursinus*) in 1955, and the publication of this work marked the first paper about an African monkey species to appear in the journal *Behaviour*. Hall undertook observations of chacma baboons in April of 1958 [DeVore and Hall, 1965], and DeVore initiated studies of olive baboons (*P. anubis*) in March of 1959 [Hall, 1968; DeVore and Hall, 1965]. Census work on yellow baboons (*P. cynocephalus*) began in September of 1959 with Washburn and DeVore [DeVore and Hall, 1965]; Kummer and Kurt began their work with hamadryas baboons (*P. hamadryas*) in Ethiopia in late 1960 [DeVore and Hall, 1965].

While groups of baboons have been studied longitudinally by a number of researchers and research teams [e.g., Altmann et al, 1985], other species of African monkeys have proven more difficult to habituate and follow. The difficulties in observing these other species, especially the forest-living monkeys, probably accounts for the relatively small number of studies, and hence, the limited amount of behavioral data, for these species. This problem was succinctly stated by Aldrich-Blake [1970], who wrote:

> With species such as baboons a troop can generally be followed throughout the day and a picture of the range can be built up by plotting daily movements over several days. In the case of the blue monkey the fragmentation of the group and the difficulties of maintaining contact with the animals over long periods render this approach impracticable, particularly since it is not always clear to which group the party of monkeys being watched belongs. (p. 87)

If groups cannot be reliably identified, the description and characterization of social behavior and its organization will likewise be hindered.

Laboratory Research

As indicated earlier in this introduction, designating a particular study as the first is difficult and risky. It is likely that such a designation would be challenged by investigators who have a different historical perspective or criterion. This review, then, is not intended to be an exact identification of the first studies that utilized African primates in studies of behavior conducted in laboratory settings, but rather a brief account of some of the early behavioral research with captive African monkeys.

Guenons. Guenons were subjects in the early studies of learning conducted by Harlow and colleagues at the Vilas Park Zoo [Harlow et al, 1932].

Harlow et al had four guenons in their original study, representing three different species. Included in this study were a subadult female mona monkey (identified as *Lasiopyga mona*), two adult male "green guenons" (identified as *Lasiopyga callitrichus*, and now designated as *Cercopithecus sabaeus*), and an immature male mustached [sic] guenon (identified as *Lasiopyga cephus*).

Mangabeys. Mangabeys, although not commonly found in captive settings, nor frequently studied in recent years (see Results), were subjects in laboratory studies in the first third of the 20th century. The first paper published in the *Journal of Comparative Psychology* that used a monkey subject was by Revesz [1925], who studied learning and generalization in two rhesus monkeys ("*Pitheci rhesi*") and a sooty mangabey ("*Cercocebus fulginosis*" [sic]). The species name of this mangabey should probably have been *fuliginosus*, which has more recently been called *Cercocebus atys* [Napier and Napier, 1967], but even more recently classified with the white-collared mangabey in the *Cercocebus torquatus* group [Napier and Napier, 1985]. Harlow et al [1932] also included a "sooty mangeby" [sic] in their delayed reaction tests. Similarly, Moss and Harlow [1947] reported using a mangabey of unspecified species in a discrimination learning study, in which they also used seven rhesus monkeys.

Baboons. Examination of the available early literature indicates that baboons were the subjects of "experimental" research in the mid-1800s, as Darwin [1872, reprinted 1965] described the facial expressions of a zoo-living anubis baboon. Darwin's work prompted Brown's [1878] experiment at the Philadelphia Zoological Garden, where he placed a snake in a mixed-species exhibit of baboons, macaques, and (likely) other primate species. The coiled, dead snake held the monkeys' attention for hours [Brown, 1878]. There was one baboon in Watson's [1908] study of imitation by monkeys, but apparently this baboon failed to perform at the level of the two rhesus and one capuchin that were also in the study. In a footnote, Watson described this baboon as being "so stupid" (p. 170) that the data from the baboon were not even presented. Hamilton's [1914] study of the sexual "tendencies" of nonhuman primates included two baboons of unspecified species. These two baboons were not chacma baboons, as the male's physical size was compared to that of chacma baboons. A mention of the male's grey color would indicate that these were hamadryas baboons, if in fact, they were baboons at all. Interestingly, Hamilton [1914] reported that the female baboon had frequently copulated with male macaques (rhesus and/or crab-eating monkeys) [see Maple et al, 1976].

Baboons that were on exhibit in zoos became popular subjects for study in the late 1920s and into the 1930s. Perhaps the best-known of these studies are those by Zuckerman [1932], who studied the colony of hamadryas baboons that lived on "Monkey Hill" at the London Zoological Garden. Included in this work by Zuckerman were his comments about breeding seasonality, based on observations he had made on wild chacma baboons in South Africa. Other comparative information presented in this 1932 volume indicates some of the other African species maintained at the London Zoological Garden.

In the United States, Harlow and colleagues, beginning in 1930, used the collection of the Vilas Park Zoo in Madison, Wisconsin, for learning studies [Harlow et al, 1932; Suomi and LeRoy, 1982]. Harlow et al's 1932 publication, Harlow's first regarding nonhuman primates [Suomi and LeRoy, 1982], included as subjects one yellow baboon, two guinea baboons, and three mandrills.

Studies of Literature Trends

Southwick and Smith's [1986] review of trends in field studies of primate ecology and behavior provides additional information about African primate research. The long-term studies of baboons mentioned above, in combination with the studies of chimpanzees and gorillas, contributed over half (501/928) of the publications dealing with African primates until the early 1970s [Southwick and Smith, 1986]. By 1975, just over half (54%; 38/71) of the publications focused on species other than baboons and great apes. The other species receiving attention were those in the genera *Cercopithecus*, *Cercocebus*, *Colobus*, and *Lemur*. Southwick and Smith's analysis showed publications on *Colobus* species gradually increasing between 1969 and 1977, at which time a gradual decline became evident. Publications on *Cercopithecus* species remained fairly constant between 1967 and 1977, peaked in 1978, and then returned to pre-1978 levels. *Cercocebus* were not considered in detail by Southwick and Smith, whereas publications on *Lemur* species showed year-to-year fluctuations after 1970, reaching a maximum number of publications in 1977.

Terry [1980] had previously analyzed the frequencies of citations regarding primate species under all headings in *Current Primate References*, rather than just those categorized as field studies of behavior and/or ecology. Terry's analysis indicated that between 1970 and 1979, the number of citations regarding four genera of Old World monkeys (*Cercopithecus*, *Presbytis*, *Colobus*, and *Theropithecus*) increased more than the average, with *Presbytis* and *Colobus* showing "growth ratios" (ratio of 1979 total to 1970 total) of more than .3 above the mean [Terry, 1980]. The rankings of "common laboratory taxa" based on the number of citations during the 1970s indicated that Old World monkey species occupied nine of the 15 positions [Terry, 1980, p. 8]. Of the African species, *P. cynocephalus* was ranked second, *C. aethiops* fourth, *Erythrocebus patas* seventh, and *P. hamadryas* ninth. The other five species were Asian macaques. For the 2-year period between July 1980 and June 1982, Erwin's [1982] similar analysis indicated that studies of *P. cynocephalus* were not as numerous, relatively, as they had been, and were ranked third; *C. aethiops* remained in fourth place. *E. patas* had also decreased relative to the others, and was in eighth place. *P. hamadryas* was ranked tenth. Like the counts done by Terry, all other ranks were occupied by species of Asian macaques.

This chapter updates and expands the analysis done by Erwin [1982], and focuses on the use of African monkeys in behavioral research during the 5-year period 1980 through 1984. The data for African monkey species are compared with data for Old World monkeys in general, and more specifically with those for Asian monkeys, particularly macaques.

METHODS

The literature trends presented here were generated from the citations contained in the monthly issues of *Current Primate References*, published by the Primate Information Center of the Regional Primate Research Center at the University of Washington. Two basic frequency counts were made. The first count was to determine the total number of citations regarding each genus or species of Old World monkey, as well as to obtain the total number of citations contained in each monthly issue. The second count was limited to those citations contained in the "Behavior" sections of these issues, with the focus again placed on Old World monkeys. Monthly totals were then combined to yield a total for each year of the 5-year period, 1980 through 1984.

Three limitations, stemming from the use of *Current Primate References* as a source of data, are placed on these frequency counts. First, the system used by the Primate Information Center to classify references must be accepted. That is, those citations that appear in the "Behavior" section must be taken to be primarily behavioral in nature. Second, the taxonomy used by the Center also must be accepted conditionally, although this taxonomy is different from those of other workers [e.g. Napier and Napier, 1967, 1985; Chiarelli, 1972; Jolly, 1973]. In the analyses presented here, several modifications of the Primate Information Center's taxonomic system were made. Specifically, Allen's swamp monkeys were included with the genus *Cercopithecus* rather than as a genus by itself (e.g. *Allenopithecus*). The

Primate Information Center includes *C. erythrogaster*, *C. erythrotis*, and *C. petaurista* as subspecies of *C. cephus;* the frequencies for these taxa have been included with *C. cephus*. Similarly, the frequencies for *C. campbelli, C. pagonias,* and *C. denti* are included with *C. mona,* as these taxa are treated as subspecies by the Primate Information Center. Genus *Mandrillus* remains under *Papio* in accordance with *CPR* usage. The third limitation is that several citations, especially those that are references to textbooks, contain mention of multiple species, thus overrepresenting the actual number of studies of a particular species. Several issues per year of *Current Primate References* contain references to published abstracts of papers presented at meetings of professional societies. These abstracts are presumed to have undergone peer and/or editorial review. These societies, along with the names of the journals that contain the abstracts, are the American Society of Primatologists (*American Journal of Primatology*), the American Association of Physical Anthropologists (*American Journal of Physical Anthropology*), and the International Primatological Society (*International Journal of Primatology* and *Anthropologica Contemporanea*). The abstracts of other professional societies are also included in *Current Primate References*, but are not generally included in the "Behavior" section.

As some comparisons were made at the genus level, the frequency data were combined across the appropriate species to produce the value for the entire genus. The following genera were included in these analyses: *Cercocebus*, *Cercopithecus*, *Erythrocebus* and *Mandrillus*, *Miopithecus*, *Macaca*, *Papio* (including *Theropithecus*), and *Presbytis*. As the base number of citations varied across years, all totals were converted to percentages to facilitate comparisons.

Chi-square tests were performed on the frequency data for the total number of citations and the total number of citations to behavioral studies for each genus to assess changes within this 5-year period. Chi-square tests for independence were done to compare the number of citations to behavioral studies to the total number of citations for all other categories (other = total − behavior), by year, for each genus. Pearson product-moment correlation coefficients (r) were calculated for these latter data as well, to provide an indication of the *direction* of the changes (if any) over the 5-year period. The magnitude of the correlation coefficient has limited utility because there were only 5 pairs of data points per correlation.

RESULTS

The total number of citations that appeared in *Current Primate References* for the 5-year period 1980 through 1984 was 22,631, of which 1,949 (8.6%)

TABLE I. Total Number of Citations, and Number of Citations in the "Behavior" Sections of *Current Primate References* (1980 Through 1984)

Category	Year 1980	1981	1982	1983	1984	5-year total
Total number or citations	3,964	4,262	4,836	4,633	4,836	22,631
Number of "behavior" citations	350	349	465	338	447	1,949
Percent of total that are "behavior" citations	8.8	8.2	9.6	7.3	9.1	8.6
Number of citations to the behavior of Old World monkeys	252	285	378	265	356	1,536
Percent of total for behavior that are citations to studies of Old World monkeys	72.0	81.6	81.3	78.4	79.6	78.8

were in the "Behavior" section. The yearly totals appear in Table I. Of these 1,949 citations, 1,536 (78.8%) referred to studies of Old World monkeys.

All Citations

The total number of citations, by genus, regarding the major genera of Old World monkeys, the percentages that these yearly totals represented, and the percentages of these citations that were in the "Behavior" sections, are presented in Table II. The genus *Macaca* accounted for nearly 60% of all citations. Among the African genera, *Papio* was cited most frequently, representing just over 14% of all citations. The other large African genus (*Cercopithecus*) accounted for 7.7% of all citations. The other genera each accounted for less than 2% of all citations.

Of the eight genera considered in these analyses, *Cercocebus*, *Colobus*, *Erythrocebus*, and *Miopithecus* did not show any significant changes in the number of citations from year to year ($p > .05$). For the African genera, the number of citations to *Papio* species showed a general increase from year to year ($\chi^2 = 11.7$, df $= 4$, $p = .02$), whereas *Cercopithecus* species had fluctuating frequencies that produced an overall difference across years ($\chi^2 = 35.5$, df $= 4$, $p < .001$). The pattern for *Macaca* was similar to that of *Papio*, showing a general increase from 1980 through 1984 ($\chi^2 = 99.2$, df $= 4$, $p < .001$), whereas the number of studies of *Presbytis* species fluctuated greatly, with a sharp peak in 1982 accounting for the significant difference across years ($\chi^2 = 28.9$, df $= 4$, $p < .001$; Table II).

In terms of the total number of citations regarding Old World monkey

TABLE II. Total Number of Citations for Old World Monkey Genera, and the Percentage of Those That Appeared in the "Behavior" Section of *Current Primate References* (1980 Through 1984)

Genus[a]	1980	1981	1982	1983	1984	Total
Cercocebus						
f	36	40	42	44	51	213
% t	0.9	0.9	0.9	0.9	1.0	0.9
% b	25.0	7.5	14.3	6.8	13.7	13.2
Cercopithecus						
f	282	331	424	324	383	1743
% t	7.1	7.8	8.8	7.0	7.8	7.7
% b	51.2	17.8	9.9	8.0	12.5	11.4
Colobus						
f	33	41	52	42	40	208
% t	0.8	1.0	1.1	0.9	0.8	0.9
% b	18.2	9.8	19.2	4.8	27.5	17.8
Erythrocebus						
f	72	64	72	68	74	350
% t	1.8	1.5	1.5	1.5	1.5	1.6
% b	1.4	6.2	6.9	5.9	5.4	5.1
Macaca						
f	2,292	2,526	2,799	2,763	2,945	13,325
% t	57.8	59.3	57.9	59.6	59.7	58.9
% b	6.3	5.9	7.8	5.2	6.5	6.3
Miopithecus						
f	22	16	28	26	17	109
% t	0.6	0.4	0.6	0.6	0.3	0.5
% b	22.7	12.5	21.4	19.2	5.9	17.4
Papio						
f	567	675	653	639	668	3,202
% t	14.3	15.8	13.5	13.8	13.5	14.2
% b	7.8	7.6	8.4	9.9	11.7	9.1
Presbytis						
f	75	61	123	68	82	409
% t	1.9	1.4	2.5	1.5	1.7	1.8
% b	26.7	21.3	29.3	28.0	19.5	25.4

[a]f, total number of citations to the genus in *Current Primate References*; % t, percentage of the total number of citations that were to the genus; % b, percentage of the genus' total that appeared in the "Behavior" sections.

TABLE III. Year by Year Listing of the Eleven Most Cited Species of Old World Monkeys in *Current Primate References* (1980 Through 1984)[a]

Species	1980	1981	1982	1983	1984	Mean rank
Macaca mulatta	1	1	1	1	1	1.0
Macaca fascicularis	2	2	2	2	2	2.0
Papio cynocephalus	3	3	3	3	3	3.0
Cercopithecus aethiops	4	4	4	4	4	4.0
Macaca nemestrina	6	5	5	5	5	5.2
Macaca arctoides	5	6	6	7	7	6.2
Macaca fuscata	7	8	7	6	6	6.8
Papio anubis	8	7	9	8	8	8.0
Erythrocebus patas	9	9	10	9	10	9.4
Macaca radiata	10	(13)	8	(12)	9	10.4
Papio papio	(11)	10	(11)	10	(11)	10.6

[a]Ranks in parentheses indicate what the ranking would have been had the listing continued beyond the first ten positions in any one year. These ranks are included in the calculation of the mean ranks.

species, the eleven most frequently cited species are listed in Table III, accompanied by the yearly rankings and the mean ranks over the entire 5-year period. For each year, the ten most cited species are indicated. Nine species appear in each of the yearly rankings, while *P. papio* and *M. radiata* alternate in the last position every other year. In each of the 5 years, the rankings of the four most frequently cited species were consistent.

For the African genera combined, approximately 10% of the total number of studies for the 5-year period were behavioral in kind. Table II contains the corresponding percentages for each genus, for each of the 5 years, and for the total time period. Of the African genera, *Colobus* has the highest percentage of behavioral studies, and *Erythrocebus* the lowest percentage.

Behavioral Studies

Considering citations of only behavioral studies of Old World monkeys, the genus *Macaca* accounted for 55.0% of the total (845/1536), followed in decreasing frequency by *Papio* and *Theropithecus* (combined; 19.0%), *Cercopithecus* (12.9%), *Presbytis* (6.8%), *Colobus* (2.4%), *Cercocebus* (1.8%), *Miopithecus* (1.2%), and *Erythrocebus* (1.2%). The yearly data and totals for these genera of Old World monkeys are presented in Table IV. Year by year differences in citation frequencies were present for the following genera: *Cercopithecus* ($\chi^2 = 22.6$, df = 4, $p < .001$), *Papio* ($\chi^2 = 11.7$, df = 4, $p = .02$), *Macaca* ($\chi^2 = 27.1$, df = 4, $p < .001$), and *Presbytis* ($\chi^2 = 15.3$, df = 4, $p = .004$). Of these, only *Papio* had greater than expected frequencies for the last 2 years (1983 and 1984). *Macaca* and *Cercopithecus* species showed year to year fluctuations, whereas *Presbytis* citations were below expected frequencies for the last two years of this analysis (see Table IV). *Cercocebus*, *Colobus*, *Erythrocebus*, and *Miopithecus* showed no overall yearly differences ($p > .05$). The year by year rankings of species within these genera, in terms of the number of citations in the "Behavior" sections, are presented in Table V.

Over the 5-year period, the number of behavior citations varied with the number of total citations for the genera *Cercocebus*, *Erythrocebus*, *Miopithecus*, and *Presbytis* (2×5 chi-square contingency tests; $p > .05$). However, the number of citations in the "Behavior" sections of *Current Primate References* was independent of the number of other citations for *Cercopithecus* ($\chi^2 = 20.6$, df = 4, $p < .001$), *Colobus* ($\chi^2 = 9.7$, df = 4, $p = .047$), *Macaca* ($\chi^2 = 17.1$, df = 4, $p < 001$), and *Papio* ($\chi^2 = 9.4$, df = 4, $p = .054$). With the latter, when the citations to the relatively infrequently studied geladas, mandrills, and drills were excluded from the other *Papio* species, the independence of the variables was more pronounced ($\chi^2 = 9.8$, df = 4, $p = .044$).

Correlation coefficients computed to estimate the direction and degree of relationship between the number of behavior citations and all other citations for each year during the 5-year period yielded negative values for *Cercocebus*, *Colobus*, and *Erythrocebus*. Inspection of these frequencies indicated that for the former two genera, studies of behavior increased relative to all other studies; for *Erythrocebus*, behavioral studies were not increasing as other types of studies were increasing in frequency. Correlation coefficients for the other genera were positive, indicating that behavioral studies were changing in frequency in the same manner or direction as other studies of that genus.

The number of behavioral citations, by year and species within the genera *Cercopithecus* and *Papio* (including *Theropithecus*), are presented in Tables VI and VII, respectively. For the genus *Cercopithecus*, studies of *C. aethiops* accounted for nearly half (45.4% of all of the citations listed), and did not differ significantly across years ($\chi^2 = 5.3$, df = 4, $p > .05$). The second most cited species within this genus was *C. mitis*, accounting for slightly over 10% of the behavioral citations. *Cercopithecus cephus* was also in this 10% range (see Table VI). *Cercopithecus ascanius* (various subspecies) and *C. mona* (various subspecies) each accounted for 8–9% of the total.

The number of behavior citations for *C. aethiops* constituted just under 8% of the total number of citations of this species, and for *C. mitis*, approxi-

TABLE IV. Number of Citations on a Yearly Basis of Genera of Old World Monkeys: Behavior Section of *Current Primate References* (1980 Through 1984)

Genus	1980	1981	1982	1983	1984	5-year total
Cercocebus						
No.	9	3	6	3	7	28
%[a]	3.6	1.0	1.6	1.1	1.9	1.8
Cercopithecus						
No.	23	58	42	26	49	198
%	9.1	20.4	11.1	9.8	13.8	12.9
Colobus						
No.	6	4	10	2	11	33
%	2.4	1.4	2.6	0.7	3.1	2.4
Erythrocebus						
No.	1	4	5	4	4	18
%	0.4	1.4	1.3	1.5	1.1	1.2
Macaca						
No.	145	149	217	143	193	847
%	57.1	52.3	57.7	54.0	53.6	55.0
Miopithecus						
No.	5	2	6	5	1	19
%	2.0	0.7	2.0	1.9	0.3	1.2
Papio[b]						
No.	44	51	55	63	78	291
%	17.5	17.9	14.6	23.8	21.9	19.0
Presbytis						
No.	20	13	36	19	16	104
%	7.9	4.6	9.5	7.2	4.5	6.8

[a]% indicates the percentage of the total for Old World monkeys.
[b]Includes *Theropithecus*.

mately 20% of the citations were to studies of behavior. The corresponding percentages for the other *Cercopithecus* species are included in Table VI.

For *Papio* species, *P. cynocephalus* received the most behavior citations (39.9%), followed by *P. anubis* (23.4%; see Table VII). Citations for *P. hamadryas* and *P. ursinus* were about equally represented over the 5-year period (14.5 and 13.1%, respectively). For none of these baboon species did the number of citations differ significantly over the 5-year period ($p > .05$) for each species without a frequency of zero for any year). All other *Papio* species (*P. papio*, *P. leucophaeus*, *P. sphinx*, and *T. gelada*) were each below 6% of the total.

TABLE V. The Ten Most Frequently Cited Species of Old World Monkeys for Behavioral Studies as Listed in *Current Primate References* (1980 Through 1984)

Rank	1980	1981	1982	1983	1984
1	*M. mulatta*	*M. mulatta*	*M. mulatta*	*M. mulatta*	*M. mulatta*
2	*M. fuscata*	*M. nemestrina*	*M. nemestrina*	*M. fuscata*	*M. fuscata*
3	*P. cynocephalus*	*P. cynocephalus*	*M. fuscata*	*M. fascicularis*	*C. aethiops*
4	*M. arctoides*	*C. aethiops*	*M. fascicularis*	*P. anubis*	*M. arctoides*
5	*M. fascicularis*	*M. fuscata*	*C. aethiops*	*M. nemestrina*	*P. cynocephalus*
6	*M. radiata*	*M. radiata*	*M. radiata*	*C. aethiops*	*P. anubis*
7	*C. aethiops*	*M. fascicularis*	*M. arctoides*	*P. hamadryas*	*M. fascicularis*
8	*M. nemestrina*	*M. arctoides*	*P. cynocephalus*	*P. cynocephalus*	*M. sylvanus*
9	*P. anubis*	*P. anubis*	*P. anubis*	*P. ursinus*	*M. nemestrina*
10	*Cercocebus* (sp.)	*P. hamadryas*	*P. ursinus*	*M. arctoides*	*M. radiata* *Colobus* (sp.)

TABLE VI. Number of Citations in *Current Primate References* to Species in the Genus *Cercopithecus* for the 5-Year Period 1980 Through 1984: Behavior

Species	1980	1981	1982	1983	1984	Total	% of genus total	% of species total that are behavioral
C. aethiops	11	20	21	15	23	90	45.4	8.0
C. ascanius	2	3	6	1	6	18	9.1	26.5
C. cephus	0	9	6	2	3	20	10.1	32.3
C. diana	0	4	1	0	3	8	4.0	17.0
C. hamlyni	0	2	0	0	0	2	1.0	13.3
C. l'hoesti	1	2	2	0	1	6	3.0	20.0
C. mitis	4	2	5	2	8	21	10.6	20.6
C. mona	1	10	0	4	2	17	8.6	18.7
C. neglectus	2	2	1	1	1	7	3.5	12.1
C. nictitans	2	2	0	1	1	6	3.0	10.5
C. nigrividus	0	0	0	0	1	1	0.5	5.0
C. wolfi	0	1	0	0	0	1	0.5	33.3
Unspecified species	0	1	0	0	0	1	0.5	2.7

The number of citations of behavioral studies for species in the genus *Macaca* are presented in Table VIII for comparative purposes. The single most studied species was *M. mulatta*, which accounted for 37.9% of the total within this genus. Overall, however, only 4.6% of the citations to *M. mulatta* were in the "Behavior" section. *Macaca fuscata*, *M. nemestrina*, and *M. fascicularis* each accounted for more than 10% of the total number of behavior citations within this genus.

TABLE VII. Number of Citations in *Current Primate References* to Species in the Genus *Papio* (Including *Theropithecus*) for the 5-Year Period 1980 Through 1984: Behavior

Species	1980	1981	1982	1983	1984	Total	% of genus total	% of species total that are behavioral
P. cynocephalus	22	23	15	9	21	90	30.9	5.8
P. anubis	9	9	11	18	21	68	23.4	14.3
P. papio	1	1	4	5	4	15	5.2	4.8
P. hamadryas	7	7	8	13	7	42	14.5	14.8
P. ursinus	4	4	11	9	10	38	13.1	16.4
P. leucophaeus	0	1	0	1	2	4	1.4	8.0
P. sphinx	0	3	1	3	2	9	3.1	10.8
T. gelada	1	2	5	5	4	17	5.8	13.8
Unspecified species	0	1	0	0	7	8	2.8	9.2

DISCUSSION

The total number of citations of studies of nonhuman primates increased over the 5-year period from 1980 through 1984. During this period, the percentages of citations regarding primate behavior fluctuated slightly, remaining between 7 and 10% of the total. The mean percentage of 8.6% is comparable to Erwin's [1981] calculation of 9.6%, which was based on a single year within this 5-year period. Erwin [1981] estimated that *field studies* of behavior constituted about 25% of the citations of studies of behavior, or approximately 2% of the total number of citations. The number of field studies of several genera were presented in Southwick and Smith's [1986] Figure 7. Estimating the 1980 and 1981 frequencies from their Figure 7, and combining those data with the frequency counts for those 2 years (data from the present chapter), the percentage of field studies for the African genera *Colobus*, *Cercopithecus*, and *Papio* were calculated. These percentages were higher than the estimate of Erwin [1981] that was calculated for all species of nonhuman primates regardless of geographic location. Field studies of *Cercopithecus* accounted for 33% of the behavior citations (27/81) in those 2 years, and 4.4% of the total number of citations (27/612). For *Colobus*, nine out of the ten behavioral citations were field studies, and represented 12.2% (9/74) of the total number of citations to that genus. Field studies of Papio accounted for 42% (40/95) of the behavioral studies, and 3.2% of the total (40/1242). It is possible that the proportion of

TABLE VIII. Number of Citations in the Behavior Section of *Current Primate References* to Species in the Genus *Macaca* for the 5-Year Period 1980 Through 1984

Species	1980	1981	1982	1983	1984	Total	% of genus total	% of species total that are behavioral
M. assamensis	1	0	2	0	0	3	0.4	4.4
M. maura	1	0	0	1	1	3	0.4	12.0
M. cyclopis	2	0	0	0	0	2	0.2	2.4
M. arctoides	17	12	17	8	21	75	8.8	11.3
M. fascicularis	15	16	24	20	20	95	11.2	3.1
M. fuscata	24	18	30	23	35	130	15.4	20.4
M. mulatta	53	55	78	63	72	321	37.9	4.6
M. nemestrina	10	24	32	17	14	97	11.4	11.5
M. nigra	1	2	2	1	0	6	0.7	6.4
M. radiata	12	17	20	7	11	67	7.9	19.4
M. silenus	1	0	3	3	2	8	0.9	10.0
M. sylvana	5	3	0	6	16	30	3.5	28.8
M. sinica	1	0	3	0	1	5	0.6	16.7
M. thibetana	0	0	0	0	0	0	0.0	0.0
Unspecified species	2	2	0	1	0	5	0.6	1.7

field studies has decreased in the past several years relative to the proportion of studies of captive nonhuman primates as new avenues for the publication of studies conducted with captive animals have opened. Thus, a decreased proportion might be an artifact of other increases rather than true decreases in frequency.

The percentage of citations of the behavior of Old World monkeys has remained relatively stable at approximately 80% of the total number of citations in the "Behavior" section of *Current Primate References.* Among the Old World monkeys, African monkeys have been disproportionately understudied in comparison to the Asian macaques in behavioral research as well as in other areas of primatology. In the 5-year period covered by the analyses presented in this chapter, *Papio* species were the most frequently listed of the African monkeys in *Current Primate References* (all topics), with *Cercopithecus* species the second most frequently listed. *Papio* species also were listed more frequently in the "Behavior" section than any other species in any African monkey genus, although several other species in the genus *Cercopithecus* had higher *proportions* of their totals listed in the "Behavior" section. Less frequently cited genera, such as *Cercocebus*, *Colobus*, and *Miopithecus*, had even higher proportions of their totals listed in the "Behavior" sections. These higher proportions of behavior citations indicate relatively little use of these species/genera in neuroscience or biomedical research.

In behavioral studies, species of macaques were consistently the two most frequently cited species, and in 1982, comprised the four most cited species. Macaque species occupied at least five of the ten positions in each of the years included in this analysis. The number of baboon species in these rankings varied between two and four, depending on the year, and only one *Cercopithecus* species (*C. aethiops*) was found consistently in these yearly rankings; *C. aethiops* was ranked at a high of third in 1984, and a low of seventh in 1980. The ratios of behavior citations of macaques to those of baboons was nearly 3:1, and for macaques to guenons, slightly greater than 4:1.

Terry's [1980] analysis of the citations that appeared during the 1970s allows for several comparisons. Terry's ranking of species commonly found in laboratory studies corresponds rather closely with those of the present analyses. Extracting only the Old World species from Terry's listings, and comparing that ranking with the mean species ranks of 5-year period analyzed here, close correspondence was found. The actual ranks of the species, however, differed in the two analyses. In both rankings, rhesus monkeys (*M. mulatta*) were the most frequently cited. Terry [1980] found that approximately 55% of all citations of laboratory studies were to rhesus monkeys. Although the present analysis did not differentiate between "common laboratory taxa" and any other taxa, citations to the genus *Macaca* represented 58.9% of the total in *Current Primate References* (Table II), and 55% of all citations to behavioral studies of Old World monkeys (Table IV). Rhesus alone accounted for 20.9% of all behavior citations during the 1980–1984 period. Terry [1980] found the common baboons (*P. cynocephalus*) to be the second most cited species for laboratory studies, with crab-eating monkeys (*M. fascicularis*) third. The 5-year period following Terry's analysis saw these latter two species undergo a reversal in terms of their ranks. In both counts, however, *C. aethiops* occupied the fourth position. Another reversal in ranks occurred that involved *M. arctoides* and *M. nemestrina*. In the 1970s, citations to *M. arctoides* were most prevalent, while in the first half of the 1980s, citations to *M. nemestrina* were slightly more prevalent than citations to *M. arctoides*. *E. patas* decreased in rank from seventh to between ninth and tenth, whereas *M. fuscata* increased in rank from eighth in the 1970s to approximately seventh in the early 1980s. In the first half of the 1980s, *M. radiata* and *P. papio* became more frequently cited, while *P. hamadryas* became less frequently cited, and no longer occupied a position in the first ten. Unfortunately, there are no readily accessible comparative data for laboratory studies of behavior alone; the above comparisons are for all citations combined.

Inspection of the data for those genera characterized by the independence of the number of behavior citations and the number of other citations revealed different patterns for the 5-year period. Frequencies for *Colobus* fluctuated yearly as to whether the number of citations to behavioral research were above or below the frequencies expected based on the 5-year totals. For the genus *Papio*, the number of behavior citations increased relative to "others" for the last 2 years of the 5-year period, whereas for the genus *Cercopithecus*, citations for behavioral studies were greater than expected relative to "other" citations for the first 4 years of this time period. In 1984, however, the number of citations to behavioral studies declined relative to other kinds of research. For *Macaca*, the general pattern was for fewer behavior citations relative to all others, and is most likely a reflection of the use of macaques in medical research. In only 1 year (1982) did the number of citations to behavior studies exceed the expected value by any appreciable amount.

These trends, although informative, should be interpreted cautiously for several reasons. First, a 5-year period may not be sufficiently long enough to detect any stable or consistent trends. Thus, these trends are short-term, and perhaps transitory. Second, publication lag and "citation lag" (length of time between publication and appearance in *Current Primate References*) could distort the frequency counts. The yearly divisions employed in the analyses contained herein were arbitrary; a study listed in January is not inherently different from one listed the previous month, although two citations appearing in this manner would be counted as having appeared in different years. This type of distortion, however, would likely only effect only those species that are infrequently listed. A third potential problem in interpreting the data presented here is that several publications (or citations) might have emanated from a single research project, which is especially likely from a long-term field study. Thus, the number of citations may not accurately represent the number of distinct studies actually conducted. Similarly, the length of field studies may result in the underrepresentation of work dealing with a particular species. The patterns of publications from field research may be more sporadic, occurring in bursts, with long interpublication intervals that reflect the lengths of the studies. Additionally, in bibliographic counts such as those presented here, each publication listed is treated equally. That is, the lengths or types of publications are not considered, whereby a brief note is counted the same as a major theoretical paper or a lengthy field report. Perhaps an analysis of the number of pages per species or genus would be a more sensitive indicator of trends in research and species selection. The limitations mentioned in the Methods section also should be taken into account when interpreting the data and trends described here.

A note about the taxonomy of guenons should be included here. The taxonomy used by the Primate Information Center differs slightly from other taxonomies. Napier and Napier [1967] classified 21 species of *Cercopithecus* monkeys into nine groups, and treated talapoins and Allen's swamp monkeys as separate species in their own respective genera. Chiarelli [1972] and Jolly [1972] included these latter two species in the genus *Cercopithecus*, and each of these authors listed 22 species in the genus. However, the 22 species were not the same. Napier and Napier's [1985] revised taxonomy, the most recent at the time of this writing, contains 20 species of *Cercopithecus* in eight groups, and still excludes talapoins and swamp monkeys from the genus. This revised taxonomy differs from their previous one in that three species are no longer considered as distinct species (*C. pygerythrus*, *C. sabaeus*, and *C. albogularis*), and two new species have been included. These new species are *C. dryas* (Dryas guenons) and *C. salango* (Zaire diana monkeys). The taxonomic system employed in the analyses presented in this chapter is similar to that used by the Primate Information Center, but included several modifications that were described in the Methods section. The differences between the Primate Information Center system, the one used here, and the other systems described above revolved around the questions of subspecies status versus species status. As the result of the combination of several subspecies, 12 species of *Cercopithecus* monkeys have been included in the assessment of species utilization, rather than the 20–22 species described by the taxonomists cited above. (It should be noted that citations to *C. preussi* were not found in the 5-year period that was analyzed.) In terms of the purposes and goals of this chapter, these differences in taxonomic status are relatively minor. The disparity in the representation of the different genera in the behavioral literature, and even the disparity among the several major species within each genus, are evident. One intended purpose of a chapter such as this was to provide the impetus to those who have access to any of these lesser known species to initiate behavioral studies, and to publish and otherwise disseminate these data in order that more complete comparative analyses of behavior can be conducted. A second purpose was to make evident that behavioral data about some of the lesser known nonhuman primate species are available in the literature. It is hoped that this, in turn, will foster comparative analyses along appropriate phylogenetic lines [see Hodos and Campbell, 1969, for elaboration]. Knowing where to locate citations of recently published data, and when those citations appeared, should facilitate the comparative approach to the study of behavior.

CONCLUSIONS

1. An analysis of the citations contained in *Current Primate References* for the 5-year period between 1980 and 1984 indicated that 8.6% of the total were to studies of behavior. Of these studies of behavior, Old World monkeys were the subjects in approximately 79% of the citations.

2. Of the Old World monkey species, macaques, baboons, and guenons were the three most cited groups, respectively. This order was reversed when the *percentage* of citations to behavioral studies was taken into account, indicating that macaque and baboon species were more frequently used in nonbehavioral research.

3. For the 5-year period, studies of the behavior of macaques accounted for 55% of the total for Old World monkeys, baboons accounted for nearly 19%, and species in the genus *Cercopithecus* accounted for nearly 13% of the total.

4. At the species level of analysis, *C. aethiops* was the most frequently cited of the genus, accounting for 45% of all studies of *Cercopithecus* behavior. *C. mitis* and *C. cephus* had the next highest percentages of 10.6 and 10.1%, respectively.

5. For *Papio* species, *P. cynocephalus* was the most cited, with 31% of the citations. The totals for *P. anubis* accounted for 23.4% of the *Papio* total.

6. The frequency of citations of studies of macaque behavior outnumbered the frequency of citations regarding baboons by 3:1, and outnumbered the frequency of citations of guenon research by approximately 4:1.

7. The relationship between the number of citations regarding behavior was independent of the number of citations of other topical areas for the genera *Colobus*, *Papio*, *Cercopithecus*, and *Macaca*. However, this independence was reflected in different patterns for each of the genera. For the other African genera, the frequencies of citations to behavioral studies and citations to all other areas of research were related.

ACKNOWLEDGMENTS

The authors gratefully acknowledge the assistance of Sonia Gales and Courtenay Smith for compiling some of these data; Charles Southwick for

Hall KRL, Goswell MJ (1964). Aspects of social learning in captive patas monkeys. Primates 5:59–70.

Hall KRL, Mayer B (1966). Hand preferences and dexterities of captive patas monkeys. Folia Primatol 4:169–185.

Hall KRL, Mayer B (1967). Social interactions in a group of captive patas monkeys (Erythrocebus patas). Folia Primatol 5:213–236.

Hall KRL, Boelkins RC, Goswell MJ (1965). Behaviour of patas monkeys, Erythrocebus patas, in captivity, with notes on the natural habitat. Folia Primatol 3:22–49.

Hamilton GV (1914). A study of sexual tendencies in monkeys and baboons. J Anim Behav 4:295–318.

Harlow HF, Uehling H, Maslow AH (1932). Comparative behavior of primates. I. Delayed reaction tests on primates from the lemur to the orang-outan. J Comp Psychol 13:313–343.

Hodos W, Campbell CBG (1969). Scala naturae: Why there is no theory in comparative psychology. Psychol Rev 76:337–350.

Jolly A (1972). "The Evolution of Primate Behavior." New York: Macmillan Company.

Lumsden WHR (1951). The night-resting habits of monkeys in a small area on the edge of the Semliki Forest, Uganda. A study in relation to the epidemiology of sylvan yellow fever. J Anim Ecol 20:11–30.

Maple TL (1979). Primate psychology in historical perspective. In Erwin J, Maple TL, Mitchell G (eds): "Captivity and Behavior: Primates in Breeding Colonies, Laboratories and Zoos." New York: Van Nostrand Reinhold Company, pp 29–58.

Maple T, Scott S, Redican WK (1976). Intergeneric sexual behavior between an adult baboon and an adult macaque. Bull Zool [illegible]:175–180.

Mitchell BL (1948). A naturalist in Nyasaland. Nyasaland Agric Q J [illegible].

Moss EM, Harlow HF (1947). The role of reward in discrimination learning in monkeys. J Comp Physiol Psychol 40:333–342.

Napier JR, Napier PH (1967). "A Handbook of Living Primates." London: Academic Press.

Napier JR, Napier PH (1985). "The Natural History of the Primates." Cambridge: The MIT Press.

Pare RV (1950). Patas and others saved their friends. Afr Wildlife 4:218–219.

Peters G (1923). Experimental study in abstraction in monkeys. J Comp Psychol [illegible]:335–343.

Southwick CH, Smith RB (1986). The growth of primate field studies. In Mitchell G, Erwin J (eds): "Comparative Primate Biology, Vol 2A: Behavior, Conservation, and Ecology." New York: Alan R. Liss, pp 73–91.

Struhsaker TT, Gartlan JS (1970). Observations on the behaviour and ecology of the patas monkey (Erythrocebus patas) in the Waza Reserve, Cameroon. J Zool 161:49–63.

Suomi SJ, LeRoy HA (1982). In memoriam: Harry F. Harlow (1905–1981). Am J Primatol 2:319–342.

Terry M (1980). Variations in the growth of literature on different primates during the 1970's. Lab Primate Newsletter 19:6–8.

Watson JB (1908). Imitation in monkeys. Psychol Bull 5:169–178.

Watson JM (1951). The wild mammals of Teso and Karamoja: VII. Uganda J 15:193–202.

Williams JB (1980). Behavioral observations of feral and free-ranging guenons and Patas (Cercopithecus): A bibliography. Seattle: Primate Information Center.

Zuckerman S (1932). "The Social Life of Monkeys and Apes." New York: Harcourt, Brace and Company.

Comparative Behavior of African Monkeys, pages 23–37

2

Sexual Differences in Early Social Development Among Captive Patas Monkeys

Kent M. Loy and James Loy

Heaton Orchard Road, West Kingston, Rhode Island 02892 (K.M.L.), and Department of Sociology and Anthropology, University of Rhode Island, Kingston, Rhode Island 02881 (J.L.)

INTRODUCTION

Twenty years have passed since K. R. L. Hall conducted the first systematic studies of the behavior of patas monkeys (*Erythrocebus patas*) [Hall, 1965, 1967; Hall, et al, 1965; Hall and Mayer, 1967]. These early studies provided a basic ethogram for the species and elucidated several important features of patas ecology and social structure, including: 1) the animals' adaptation to an extremely terrestrial way of life; 2) the occurrence of unimale, multifemale breeding groups and all-male groups; and 3) the respective contributions of the adult male and adult females to the structure of the breeding group. With regard to the last feature, it was found that the core of a patas heterosexual group is formed by the adult females, while the adult male occupies a socially peripheral position. This arrangement has since been confirmed by several other investigators [Rowell and Hartwell, 1978; Kaplan and Zucker, 1980; Loy, 1981].

Additional information about the breeding and social systems of patas monkeys was provided by field studies from the 1970s [Struhsaker and Gartlan, 1970; Gartlan and Gartlan, 1973; Gartlan, 1975]. Of particular interest was the report that all-male patas groups occasionally contain very young animals. In one case, an animal estimated to be 18 months of age [prepubertal by some two years; Rowell, 1977] was seen in an all-male unit [Gartlan, 1975]. Observations such as these prompted numerous questions concerning the life histories of patas monkeys and the ontogeny of their various social groups. For example, what factors influence social development? What developmental paths do males and females take en route to their adult social roles? When and why does a male leave his natal group?

An opportunity to try to address these questions came in 1972. Early in that year, a lengthy study was begun of the behavior of a captive breeding group of patas monkeys held at the Caribbean Primate Research Center. As part of that investigation, one observer (KML) made regular, and the other (JL) occasional, observations of early social development. This paper reports the findings on the social and agonistic interactions of patas infants during their first 10 months of life.

METHODS

Between November 1971 and June 1972, a patas breeding group was formed at the La Parguera colony of the Caribbean Primate Research Center (Puerto Rico). The early history of the group is complex, and details have been given in another publication [Loy, 1981]. It is sufficient to note here that by mid-June 1972 the group consisted of one adult male, eight adult females, and three juveniles (Fig. 1). To the best of the authors' knowledge, all of the animals had been wild-born in Africa. Whether any of the monkeys were related to or familiar with each other prior to their arrival at La Parguera was unknown. The monkeys were introduced into a .2-hectare outdoor corral and maintained on Wayne Monkey Diet, fruit, and water.

Breeding started soon after the group was formed, and the first infants were born in the summer of 1972. Intensive observations on the infants began on August 25, 1972 and continued until June 1, 1974. Infants were focal sampled for 10-min periods on an average of 3 days each week. Observations were made from a tower adjacent to one wall of the corral. Most samples were taken between 0900–1400 hr, but some were taken as early as 0730 or as late as 1500 hr. During the study, 14 infants (six males and eight females) were observed for all or part of their first 10 months of life (Table I). A total of 1,068 focal samples was collected.

Early in the study, long-hand prose notes were made on the infants' behaviors, and the durations of selected interactions were timed with a stopwatch. After February 5, 1973, all observations were tape recorded. Interactions were later tabulated and timed from the recording.

During the focal samples, the following behaviors were recorded.

1. Body contact (sitting touching). Generally, monkeys in body contact were relaxed and sitting quietly; infants often snuggled against their mother's chests. Other behaviors that involved physical contact, such as grooming, were not included in this category.

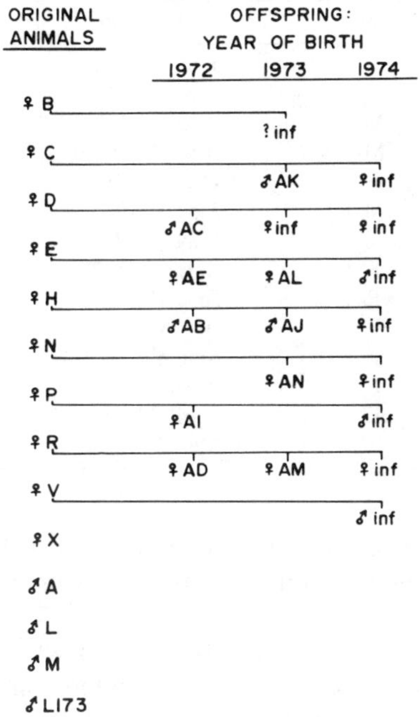

Fig. 1. Composition of the study group. Original animals were full adults when introduced into the group, with the exceptions of juveniles V, X, and M, and young adult L. Males A and L were removed from the group on June 5, 1972, leaving L173 as the single fully adult male during the remainder of the study. The two unnamed 1973 infants died shortly after birth. None of the 1974 infants had been given formal names when the study ended.

2. Allogrooming. This pattern (hereafter referred to simply as grooming) involved one animal searching through the pelage of another and removing bits of dirt, scabs, etc. Self-grooming (autogrooming) was recorded, but is not reported here.

3. Social play. This pattern consisted of wrestling, chasing, hopping, etc. [Hall et al, 1965].

4. Fighting. A fight was defined as an attack of any intensity followed by a flight of any intensity [Sade, 1967]. Attack patterns included staring, open-mouth threats; lunges, charges, and chases; grabs and slaps; and bites. The most commonly observed flight patterns included cowering and actually retreating from a threat/attack. The winner of a fight was defined as the animal whose threats/attacks caused the other monkey (the loser) to flee. "Probable wins" were sometimes scored when the observer was relatively

TABLE I. Infants Studied and Months Observed

Infant	Sex	Birthdate	Months sampled
AE	F	June 28, 1972	2–10
AD	F	July 7, 1972	2–10
AB	M	Aug. 14, 1972	1–10
AC	M	Aug. 30, 1972	1–10
AI	F	Dec. 25, 1972	1–10
AJ	M	May 1, 1973	1–10
AK	M	May 1, 1973	1–10
AL	F	May 29, 1973	1–10
AN	F	June 18, 1973	1–10
AM	F	July 5, 1973	1–10
P's 1974 infant	M	March 12, 1974	1–3
H's 1974 infant	F	April 17, 1974	1–2
C's 1974 infant	F	April 30, 1974	1–2
V's 1974 infant	M	May 2, 1974	1

certain that a win had occurred, but the entire fight had not been observed. Dominance was measured mainly in terms of winning and losing dyadic fights.

5. Displacements. A displacement was scored when one animal moved away in response to the approach of another, without any overt threat by the approaching monkey. Like winning and losing fights, this behavior was a reliable indicator of dominance.

Statistical analyses involved grouping the behavioral data into 4-week or monthly "age blocks" and then comparing observed values with expected values using one-sample Chi-square or Kolmogorov-Smirnov (K-S) tests. The null hypothesis for all tests was that males and females behaved similarly (i.e., showed equal amounts or frequencies of the behavior under consideration). When necessary, expected values were corrected for unequal focal sampling of the sexes. The alpha level for all tests was 0.05.

RESULTS

Affiliative Interactions

Infants and mothers. Three measures were used to document the complex, multi-faceted mother-infant relationship and its changes as the infant matures: 1) time spent in contact during each sample; 2) frequency of body contact; and 3) grooming between mother and infant. Both grooming and body contact are widely regarded as behaviors that reflect affiliative social relations [Sade, 1965; Berman, 1982].

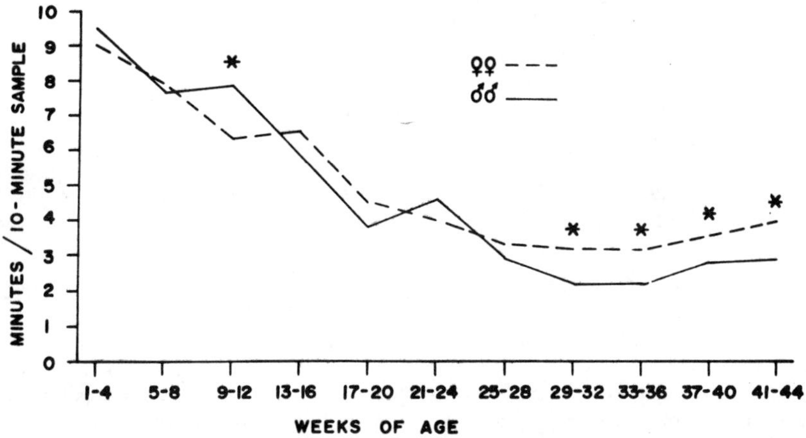

Fig. 2. Mother-infant body contact: amount. The mean time in body contact with mother during a 10-min focal sample is shown for male and female infants. Asterisks indicate significant differences from expected contact values.

With regard to the first measure, male and female infants spent about equal amounts of time in contact with their mothers during the first 28 weeks of life (Fig. 2). An exception to this generalization was the period from weeks 9–12, when females showed less, and males more, contact than expected ($X^2 = 7.99$, df = 1, $p < 0.01$). The sexes showed clear differences in maternal contact time after about the 28th week of life. Females spent more time than expected, and males less time than expected, in contact with mother between weeks 29–44 (weeks 29–32: $X^2 = 5.96$, df = 1, $p < 0.02$; weeks 33–36: $X^2 = 8.25$, df = 1, $p < 0.01$; weeks 37–40: $X^2 = 4.74$, df = 1, $p < 0.05$; weeks 41–44: $X^2 = 8.88$, df = 1, $p < 0.01$) (Fig. 2).

An analysis of the data on the frequency of contact with mother showed significant departures from expected values during month 2 ($X^2 = 9.61$, df = 1, $p < 0.01$), month 4 ($X^2 = 4.35$, df = 1, $p < 0.05$), month 8 ($X^2 = 11.85$, df = 1, $p < 0.001$), and month 10 ($X^2 = 5.38$, df = 1, $p < 0.05$) (Fig. 3). Females showed more frequent maternal contact than males in all of these months, with the exception of month 4. Both body contact measures (amount and frequency) showed decreasing interactions between mothers and both male and female infants as the infants grew older.

Mothers groomed male and female infants about equally often during the first 4 months of the infants' lives (Fig. 4). Between months 5–9, however, mothers groomed their infant daughters more, and their sons less, than expected (month 5: $X^2 = 7.28$, df = 1, $p < 0.01$; month 7: $X^2 = 6.69$,

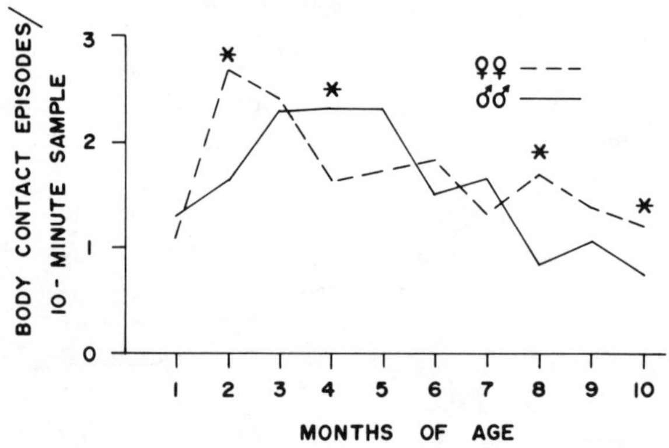

Fig. 3. Mother-infant body contact: frequency. The mean number of contact episodes per 10-min focal sample is shown for male and female infants. Asterisks indicate significant differences from expected values.

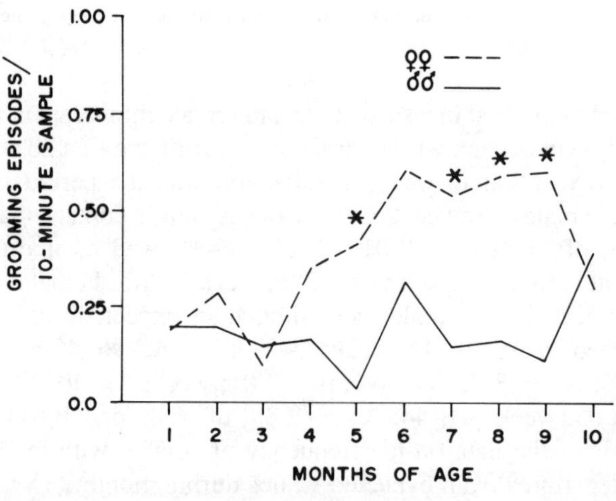

Fig. 4. Mother-to-infant grooming. The mean frequency of grooming episodes per 10-min focal sample is shown for male and female infants. Asterisks indicate significant differences from expected values.

df = 1, $p < 0.01$; month 8: $X^2 = 8.64$, df = 1, $p < 0.01$; month 9: $X^2 = 16.01$, df = 1, $p < 0.001$) (Fig. 4). Maternal grooming of daughters increased as the youngsters matured. In contrast, mothers groomed their infant sons at a relatively uniform rate throughout the first 10 months of life.

Instances of an infant grooming its mother were relatively rare. Although on one occasion a 5-month-old male was seen grooming his mother, all other infant-to-mother grooming episodes (n = 16) involved infants between 8–10 months of age. When infant-to-mother grooming was tested across the entire 10-month period, it was found that females groomed more, and males less, than expected ($X^2 = 3.88$, df = 1, $p < 0.05$).

Play was a form of friendly interaction shown by the infants as early as the first or second month of life. It did not appear, however, to be an important form of mother-infant interaction. At no time during their first 10 months of life were either female or male infants seen to engage in clearly reciprocal play with their mothers.

In summary, it appeared that there were only minor differences between males and females in their interactions with their mothers during the first 4 months of life. Between the fifth and tenth months of life, however, daughters were more often in body contact with their mothers or engaged in grooming with their mothers than were sons. It is suggested, therefore, that between 5–10 months of infant age, the mother-daughter bond becomes significantly stronger than the bond between mother and son.

Infants and unrelated adult females. Early social development is affected not only by the infant's mother, of course, but also by other groupmates. In a normally constituted patas group, it is possible for an infant to interact with the adult male, related and unrelated adult females, and related and unrelated immature monkeys of both sexes. The corral group was formed just a few months before the infant study began, using monkeys who were assumed to be strangers to one another. Therefore, certain categories of groupmates, including related adult females and related immatures of both sexes, were either missing or under-represented within the group and their effects on infant development could not be ascertained. The infants' social partners, other than their mothers, were limited, for all practical purposes, to unrelated adult females, the adult male, and unrelated infants and juveniles of both sexes.

Throughout the first 10 months of life, female infants tended to be involved in body contact with unrelated adult females more frequently than were male infants (Fig. 5). There were significant departures from expected contact frequencies during months 2, 3, and 10, and in all three instances females participated in more, and males in less, body contact than expected (month 2: $X^2 = 7.08$, df = 1, $p < 0.01$; month 3: $X^2 = 7.92$, df = 1, $p < 0.01$; month 10: $X^2 = 7.73$, df = 1, $p < 0.01$). It seems reasonable to assume that the adult females initiated most of the observed body contact episodes, especially those involving very young and relatively immobile infants [Chism, 1980]. These data suggested, therefore, that unrelated adult females were more interested in body contact with female infants than with

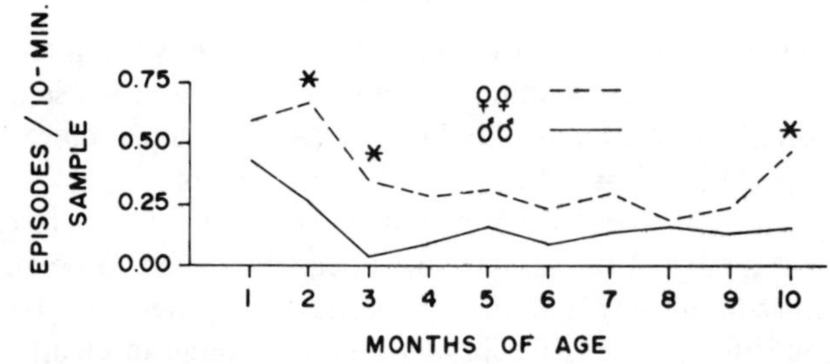

Fig. 5. Body contact between infants and unrelated adult females. The mean frequency of contact per 10-min focal sample is shown for female and male infants. Asterisks indicate significant differences from expected values.

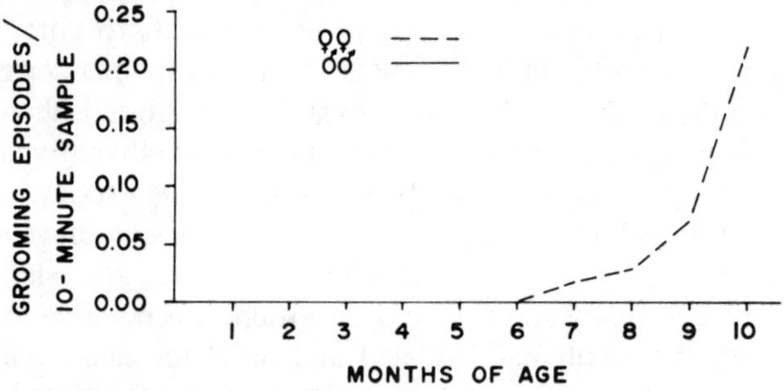

Fig. 6. Infant-to-unrelated adult female grooming. The mean frequency of grooming episodes per 10-min focal sample is shown. Male infants were not seen to groom unrelated adult females.

male infants, and that this interest was displayed very early in the infants' lives.

Unrelated adult females also groomed female infants more often than they groomed infant males. This was shown by an analysis of data summed over the entire 10-month period. Female infants were found to receive more, and male infants to receive less, grooming from unrelated adult females than expected ($X^2 = 12.54$, df= 1, $p < 0.001$).

As early as their seventh month of life, female infants began to groom unrelated adult females. The frequency of such grooming increased sharply between months 7–10 of the infants' lives (Fig. 6). Male infants were not observed grooming unrelated adult females during the study.

In summary, all measures of affiliative behaviors indicated that unrelated adult females were more attracted to female infants than to male infants. Early interactions between unrelated adult females and infant females were probably initiated by the adults. Beginning at approximately 7 months of age, the young females actively returned the adults' interests, as shown by infant-to-adult grooming episodes. Evidently, female infants were being integrated more actively into the core of the group than were male infants.

Infants and the adult male. Instances of body contact between the adult male and infants were rare. Only 15 such episodes were observed during the study. When the data were summed across the first 10 months of the infants' lives, it was found that male and female infants were in body contact with the adult male at about equal frequencies ($X^2 = 0.22$, df = 1, $p > 0.50$).

Grooming between infants and the adult male was likewise a rare occurrence. Only four instances of infants grooming the male were observed. All of these episodes featured male infants as the groomers. Female infants were not seen to groom the adult male, a finding that recalled the failure of infant males to groom unrelated adult females. The adult male was not observed grooming infants of either sex.

Two play episodes between the adult male and infants were seen. One episode involved a male infant and the other involved an infant female.

Overall, neither sex of infants appeared to be closely bonded to the adult male of the group. There were indications, however, that the infant males were more attracted to the mature male than were female infants.

Infants and unrelated immature monkeys. As early as the first month of life, infants interacted relatively frequently with unrelated agemates and juveniles. Across months 1–10, female infants were found to show more frequent body contact, and male infants less frequent contact, with unrelated immature monkeys than expected ($X^2 = 28.67$, df = 1, $p < 0.001$). Female infants also groomed unrelated immatures more frequently than expected, while male infants groomed less than expected ($X^2 = 10.30$, df = 1, $p < 0.01$). The two sexes of infants received grooming from unrelated immatures at about equal frequencies during the first 10 months of life.

Play with unrelated immatures began during the first 3 months of an infant's life and increased steadily with age. Female infants showed a decline in play with unrelated juveniles and infants after the seventh month of life, while males showed a similar decline after the eighth month. Data from ten infants (six females and four males) at 5 months of age and at 10 months of age were analyzed for evidence of sex preferences in play partners. Each animal's play frequencies with same-sex and opposite-sex partners were tested (with Chi-square and K-S tests) against frequencies expected under the null hypothesis of no sex preference among observed play partners. The results suggested that female infants had no sex preferences at either age. In contrast, by 10 months of age, three of four male infants appeared to prefer male partners (male AB: $X^2 = 4.32$, df = 1, $p < 0.05$; male AK: $X^2 = 14.34$, df = 1, $p < 0.001$; male AJ: $X^2 = 18.83$, df = 1, $p < 0.001$).

Agonistic Interactions and Dominance

The infants in the group started fighting with their groupmates as early as the fifth month of life. By observing the winners and losers of dyadic fights, we were able to determine the dominance relations of not only the infants, but of the adult animals as well. The adult females were found to be organized in a stable, linear dominance hierarchy [Loy, 1981]. Following this discovery, further analyses were done to see whether rank of infants among agemates was correlated with maternal rank, as has been reported for other species of monkeys [*Macaca fuscata*: Kawamura, 1965; Koyama, 1967; *Macaca mulatta*: Sade, 1967; Loy and Loy, 1974; *Papio cynocephalus*: Cheney, 1977; Lee and Oliver, 1979].

Dominance relations within the 1972 and 1973 birth crops were analyzed by plotting fights and displacements onto matrices according to the identities of the winners and losers (Figs. 7 and 8). For comparison with maternal relations, infants were arranged according to their mother's rank among the adult females. Infant female AI was difficult to assign to either birth crop since her birthdate fell almost exactly halfway between the two. Rather than artificially simplify the data, AI was analyzed twice, first with the 1972 group and then with the infants from 1973.

With the exception of AI, the dominance relations among the 1972 infants were linear and paralleled exactly their mother's relations (Fig. 7). AI was dominated to some extent by all four of the other 1972 infants, possibly because she was several months younger. This surmise may be incorrect, however, since birth order was not correlated with rank among infants in either year and since being older did not work to AI's advantage in 1973. In contrast to the previous year, AI's relations within the 1973 birth crop were accurately predicted by her mother's rank among the adult females (Fig. 8). In this birth crop, dominance relations among the infants were perfectly linear and in complete agreement with maternal relations.

One other possible determinant of infant rank was analyzed, namely sex. This variable was not found to be predictive. The data suggested, therefore, that the dominance ranks of patas infants with regard to their agemates are determined primarily by the dominance ranks of the youngsters' mothers.

DISCUSSION

The developmental pattern shown by the infants at La Parguera was very similar to that observed by Chism [1980] in the patas colony at the University of California, Berkeley. Chism noted that 1-month-old infants were in contact with their mothers in over 80% of half-min observation periods. By

WINNER \ LOSER	E/AE	D/AC	P/AI	R/AD	H/AB
E	—	15	22	18	17
AE		9·5·0	4·4·0	1·0·1	6·2·0
D		—	10	14	15
AC			3·0·0	6·1·2	5·6·0
P			—	26	15
AI				1·0·0	0·0·1
R			—	—	15
AD			0·1·1		6·4·0
H			—		—
AB			3·0·0		

Fig. 7. The dominance relations among mothers and among infants during 1972. Mothers are listed in large letters and infants in smaller letters. Each mother is listed beside her infant across the top of the matrix and immediately above her infant along the left margin. The large numbers indicate fights between mothers from July 16, 1972 through June 30, 1973. The smaller numbers indicate, in order, fights, displacements, and "probable" fights from the infants' first 10 months. Due to different birthdates, some of the infants' data reflect interactions with opponents older than 10 months (this is especially true for AI).

6 months of age, mothers and infants were making contact in less than 20% of the half-min periods. In the La Parguera group, infants from 1–4 weeks of age spent just over 90% of a typical 10-min sample in contact with their mothers. This figure had dropped to 30–40% of an average sample by the time the infants were 25 weeks of age (Fig. 2).

Chism [1980] noted that at Berkeley the frequency of mother-to-infant grooming was fairly constant over most of the first year of life. She did not separate the data for infant males and females. At La Parguera, the frequency of mothers grooming sons was relatively constant over the first 10 months of life. Mother-to-daughter grooming, however, showed a sharp increase with increasing infant age and occurred almost twice as often between months 6–10 as between months 1–5 (Fig. 4).

In his report on the behavior of wild patas monkeys, Hall [1965] noted that mothers did not allow other females to handle and carry their infants. Subsequent studies, however, including the present one, have shown that infant transfer (i.e., alloparenting) is a common behavior for patas. Among the Berkeley patas, first transfers usually occurred during the second week of life, and the majority of all transfers were accomplished by adult females.

LOSER

WINNER	E/AL	C/AK	P/AI	R/AM	N/AN	H/AJ
E AL	___	7 4·0·1	12 0·2·1	3 14·4·1	3 6·1·0	10 4·4·0
C AK		___	20 0·1·1	6 4·3·1	11 1·3·0	10 1·1·1
P AI			___	11 0·0·1	12 2·3·0	14 1·0·0
R AM				___	12 3·1·2	7 4·1·0
N AN					___	24 0·1·0
H AJ						___

Fig. 8. The dominance relations among mothers and among infants during 1973. Mothers are listed in large letters and infants in smaller letters. Each mother is listed beside her infant across the top of the matrix and immediately above her infant along the left margin. The large numbers indicate fights between mothers from May 1, 1973 through May 31, 1974. The smaller numbers indicate, in order, fights, displacements, and "probable" fights from the infants' first 10 months. Due to different birthdates, some of the infants' data reflect interactions with opponents older than 10 months. The infants' data also include a few "probable displacements": AK over AI once, AI over AN once, AM over AN once, and AN over AJ once.

The adult male of the Berkeley colony was not involved in taking infants [Chism, 1978, 1980]. Within the La Parguera group, many infant transfers involved adult females (Fig. 5). Toward the end of the study, juvenile sisters became frequent caretakers. The adult male of our group was not an active caretaker. In an all-female group living in a laboratory gang-cage at the University of Rhode Island, first transfers occurred at a median age of 20 days, and both adult and juvenile female patas were observed to be active allomothers [Moisen, 1981]. Finally, among the free-ranging patas observed by Zucker and Kaplan [1981] at La Parguera, alloparenting was a frequent occurrence, involved primarily adult females and juveniles (sexes unspecified), and was not displayed by the adult male of the group.

Alloparenting has been ascribed a variety of functions and effects. Zucker and Kaplan [1981] suggested that it could facilitate infant rescue in emergency situations. Moisen [1981] suggested that it could promote infant independence. Chism [1978, 1980] observed that patas yearlings often continue to be bonded closely to those animals who were their caretakers during infancy, and that former caretakers often hold and groom yearlings rejected by their mothers, and defend yearlings against threats from other monkeys.

One additional effect of alloparenting can be suggested from the present study. The data indicated that adult allomothers preferred female infants over male infants during the first few months of life (Fig. 5). Moisen [1981] made a similar finding in her laboratory study. A pattern of differential alloparenting such as this could result in different integration patterns for infant males and females. Young females would rapidly become part of the group's core, while males would lag behind. Differential alloparenting, therefore, could be one of several variables that combine to move very young males and females toward different adult social roles.

By late in their first year of life, female infants appear to initiate actively social interactions with adult females other than their mothers (Figs. 5 and 6). As new infants are born, these now juvenile females become frequent allomothers, "baby sitting" new siblings and touching, grooming, and holding the infants of their former caretakers [Chism, 1980]. In contrast, male infants have little to do with unrelated adult females, show a preference for male playmates, and show just slightly more interest in the adult male than do female infants. Further, infant males are not groomed as often by their mothers as are female infants (Fig. 4), and as yearlings, males are not used as "baby sitters" by their mothers [Chism, 1980]. Thus, even the intragenealogical bonds of young males may not be as strong as those of like-aged females.

Three interrelated factors, therefore, seem to start young patas males toward peripheralization and ultimate emigration from the natal group: 1) the development of relatively weak bonds with unrelated adult females who, along with related females, form the core of the group; 2) the development of relatively weak intragenealogical bonds; and 3) the development of a degree of unisexual attraction [Gartlan, 1975].

In closing, a word needs to be said about the demonstrated correlation between mothers' dominance ranks and the dominance relations among immature patas monkeys. Clearly, such a system stands to be of greatest importance to the nonmigrating sex; in this case, the females. Female rank has been suggested to be correlated with differential reproductive success in macaques [Drickamer, 1974] and the same could be true for patas, although at present supporting data are lacking [Loy, 1981]. For a patas male, it seems likely that the most important benefit of "inheriting" maternal rank may be to improve his chances of survival to emigration age. Whether maternally-determined rank affects a male's chances of taking over the breeder position in his natal group [Hall, 1967], or his existence within an all-male group, is unknown.

CONCLUSIONS

From the preceding information, the following suggestions about the early social development of patas monkeys are offered:

1. Male and female infants show about equal degrees of bonding with their mothers between 1–4 months of age. Between 5–10 months of age, however, the mother-daughter bond becomes significantly stronger than the mother-son bond.
2. Female infants receive more allomothering from unrelated adult females than do male infants. By the end of the first 10 months of life, female infants are more closely bonded to unrelated adult females than are infant males. This amounts to older female infants being more completely integrated into the natal group than males.
3. Male infants not only fail to become well integrated into the female core of the natal group, they develop a degree of unisexual attraction. This is shown by play partner preferences and occasional grooming of the adult male.
4. Three factors may contribute to the process of male peripheralization and ultimate emigration: a) the development of relatively weak bonds with unrelated adult females; b) the development of relatively weak intragenealogical bonds; and c) the development of a degree of unisexual attraction.
5. Dominance ranks among infant patas monkeys are correlated with maternal ranks.

ACKNOWLEDGMENTS

The authors wish to thank the staff of the Caribbean Primate Research Center for assistance during the study. At the time of the study, the CPRC was supported by contract NIH-DRR-71-2003 from the Division of Research Resources, National Institutes of Health. The authors appreciate the comments of Dr. Janice Chism, who reviewed the manuscript.

REFERENCES

Berman CM (1982): The ontogeny of social relationships with group companions among free-ranging infant rhesus monkeys I. Social networks and differentiation. Anim Behav 30:149–162.

Cheney DL (1977): The acquisition of rank and the development of reciprocal alliances among free-ranging immature baboons. Behav Ecol Sociobiol 2:303–318.

Chism J (1978): Relationships between patas infants and group members other than the mother. In Chivers DJ, Herbert J (eds): "Recent Advances in Primatology." London: Academic Press, Vol. 1, pp 173–176.

Chism J (1980): A comparison of development and mother-infant relations in patas and talapoin monkeys. PhD Dissertation, University of California, Berkeley.

Drickamer LC (1974): A ten year summary of reproductive data for free ranging *Macaca mulatta*. Folia Primatol 21:61–80.

Gartlan JS (1975): Adaptive aspects of social structure in *Erythrocebus patas*. In Kondo S, Kawai M, Ehara A, Kawamura S (eds): "Proceedings From the Symposia of the Fifth Congress of the International Primatological Society." Tokyo: Japan Science Press, pp 161–171.

Gartlan JS, Gartlan SC (1973): Quelques observations sur les groupes exclusivement males chez *Erythrocebus patas*. Ann Faculte Sci Cameroun 12:121–144.

Hall KRL (1965): Behaviour and ecology of the wild patas monkey, *Erythrocebus patas*, in Uganda. J Zool 148:15–87.

Hall KRL (1967): Social interactions of the adult male and adult females of a patas monkey group. In Altmann SA (ed): "Social Communication Among Primates." Chicago: University of Chicago Press, pp 261–280.

Hall KRL, Boelkins RC, Goswell MJ (1965): Behavior of patas monkeys, *Erythrocebus patas*, in captivity, with notes on the natural habitat. Folia Primatol 3:22–49.

Hall KRL, Mayer B (1967): Social interactions in a group of captive patas monkeys (*Erythrocebus patas*). Folia Primatol 5:213–236.

Kaplan JR, Zucker E (1980): Social organization in a group of free-ranging patas monkeys. Folia Primatol 34:196–213.

Kawamura S (1965): Matriarchal social ranks in the Minoo-B troop. In Altmann SA (ed): "Japanese Monkeys: A Collection of Translations." Atlanta: published by the editor, pp 105–112.

Koyama N (1967): On dominance rank and kinship of a wild Japanese monkey troop in Arashiyama. Primates 8:189–216.

Lee PC, Oliver JI (1979): Competition, dominance and the acquisition of rank in juvenile yellow baboons (*Papio cynocephalus*). Anim Behav 27:576–585.

Loy J (1981): The reproductive and heterosexual behaviours of adult patas monkeys in captivity. Anim Behav 29:714–726.

Loy J, Loy K (1974): Behavior of an all-juvenile group of rhesus monkeys. Am J Phys Anthropol 40:83–96.

Moisen DW (1981): Allomothering and infant independence in captive patas monkeys (*Erythrocebus patas*). MS Thesis, University of Rhode Island, Kingston.

Rowell TE (1977): Variation in age at puberty in monkeys. Folia Primatol 27:284–296.

Rowell TE, Hartwell KM (1978): The interaction of behavior and reproductive cycles in patas monkeys. Behav Biol 24:141–167.

Sade DS (1965): Some aspects of parent-offspring and sibling relations in a group of rhesus monkeys, with a discussion of grooming. Am J Phys Anthropol 23:1–18.

Sade DS (1967): Determinants of dominance in a group of free-ranging rhesus monkeys. In Altmann SA (ed): "Social Communication Among Primates." Chicago: University of Chicago Press, pp 99–114.

Comparative Behavior of African Monkeys, pages 39–65

3

Behavioral Development and Proximity Patterns in Captive DeBrazza's Monkeys

Barbara C. Kirkevold and Carolyn M. Crockett

Department of Psychology, University of Washington, and Woodland Park Zoological Gardens, Seattle, Washington 98195

INTRODUCTION

The DeBrazza's monkey (*Cercopithecus neglectus*) is one member of a diverse genus of African monkeys. Taxonomic assignments within this complex genus vary from source to source, but all authors agree upon the classification of *C. neglectus* into one monotypic species [Pocock, 1907; Schwartz, 1928; Hill, 1966; Napier and Napier, 1967]. Although widely distributed in central Africa [12°E–35°E, 7°N–11°S; Napier and Napier, 1967; Wolfheim, 1983], few field observations have been published [Brown and Urban, 1969; Quris, 1976; Gautier-Hion, 1978; Gautier-Hion and Gautier, 1978; Brennan, 1985]. The typical habitat for DeBrazza's monkeys is swamp forest, usually in proximity to rivers or streams. In Gabon, DeBrazza's primarily coexist with *Cercocebus galeritus* although some groups are sympatric with *Cercopithecus nictitans*, *C. pogonias*, *C. cephus*, *Cercocebus albigena*, *Miopithecus talapoin*, and *Colobus abyssinicus* [Quris, 1976; Gautier-Hion and Gautier, 1978]. In southwestern Ethiopia, DeBrazza's are sympatric with *C. mitis* and *C. aethiops* [Brown and Urban, 1969]. Unlike most *Cercopithecus* species, DeBrazza's monkeys rarely form polyspecific associations, and in many respects, are unique members of the genus [Quris, 1976; Gautier-Hion and Gautier, 1978]. Most notably, they form very small groups of about three to six individuals composed of one adult male, usually one and sometimes two adult females, and their offspring [Quris, 1976; Gautier and Gautier-Hion, 1977; Gautier-Hion and Gautier, 1978; Gautier-Hion, 1978; Brennan, 1985]. One study of stomach contents found their diet to be 74% fruits and seeds, 9% leaves, 5% insect and other animal matter, and 12% flowers, mushrooms, and earth [Gautier-Hion, 1978]. In riverine habitat DeBrazza's typically spend more than 50% of the time below 10 meters or on the ground, but they become more arboreal in primary forest [Quris, 1976; Gautier-Hion and Gautier, 1978; Gautier-Hion, 1978]. Small group size, quiet locomotion, freezing behavior, and the adult male's large size and vocal display have been attributed in part to antipredator strategies [Gautier-Hion and Gautier, 1978].

The vigilance and well-developed antipredator responses of DeBrazza's monkeys have made field observations of their social behavior extremely difficult. Thus, data from captive groups can contribute to our understanding of this species. Past studies of captive DeBrazza's have focused on infant development [Chalmers, 1972; Stevenson, 1973], comparative social interactions [Chalmers, 1973], general behavior patterns [Mörike, 1976; Oswald and Lockard, 1980], communication [Gautier and Gautier-Hion, 1977], and reproduction [Rowell and Richards, 1979]. The present study focused on a group of DeBrazza's housed at Seattle's Woodland Park Zoological Gardens. The monkeys had reproduced successfully for several years prior to the onset of observations, and had not been disrupted nor handled. Since infants were born every year, this group was ideal for studying behavioral development. Although most behavioral observations focused on the infants, additional data were collected to characterize the other age-sex classes and general social organization of the troop.

METHODS

Subjects

The study group consisted of an adult male (MD), two adult females (FB and FS), and their nine offspring: two 4-year-old subadult females (AB and AS), two 3-year-old juvenile males (B1 and B2), two 2-year-old juvenile males (C1 and C2), and three infants who were 10 months old (male DS), 7 weeks old (female EB), and 1 week old (male ES) at the onset of observations (Tables I and II).

Male MD and females FB and FS were acquired as adults and were apparently wild-caught. All offspring born prior to 1972 had been sold or died by 1972. Thereafter, the group was stable and never handled until 1977 when the monkeys were captured, sexed, and marked. Beginning in 1971 both adult females gave birth annually, and with the exception of FB's 1975 stillborn infant, female FB always preceded female FS (Table I). Immatures AB, B1 and C1 were slightly larger than their agemates AS, B2, and C2, and

TABLE I. Subjects and History of Woodland Park Zoo DeBrazza's Group

ID	Zoo no.	Sex	Date born	Mother	History
FB	WD-1	F	Before '62	—	Acquired as adult, July '66; died 1981
MD	WD-2	M	Before '62	—	Acquired as adult, July '66; sold 1981
	None	—	Apr. 1, '67	FB	Sold 1969
	WD-3	—	June 15, '68	FB	Sold 1972
	None	—	'69	FB	Died 1970?
	None	—	Mar. 25, '70	FB	Died 1970
FS	WD-4	F	Before '66	—	Acquired as adult, May 70; removed April '77 (injured)
	WD-5	—	May 4, '71	FB	Sold/died?
	WD-6	—	July 14, '71	FS	Sold 1972
AB	WD-7	F	June 12, '72	FB	Died 1978
AS	WD-8	F	July 25, '72	FS	Died 1981
B1	WD-9	M	July 7, '73	FB	Sold 1978
B2	WD-10	M	Sept. 3, '73	FS	Sold 1977 (after study)
C1	WD-11	M	Aug. 27, '74	FB	Sold 1980
C2	WD-12	M	Sept. 28, '74	FS	Sold 1984
DS	WD-13	M	Sept 22, '75	FS	Sold 1985
	WD-17	—	Oct. 4, '75	FB	Stillborn
EB	WD-14	F	June 4, '76	FB	Died 1980
ES	WD-15	M	Sept. 4, '76	FS	Died 1979
	WD-16	M	July 25, '77	FB	Sold 1983
	WD-18	M	July 5, '78	FB	Sold 1984
	WD-19	F	Oct. 22, '79	FB	On display 1985
	WD-20	—	Nov. 14, '80	FB	Died 1981

TABLE II. Infant Subjects and Ages; Total Sample Durations per Subject

	Infant ages in weeks		
Dates	ES male	EB female	DS male
7/17–9/13/76		7–10	43–46
9/9–11/30/76	1–12	14–25	50–61
4/28–7/28/77[a]	33–46	47–60	83–96
Total observation (in hours)	12.33	22.68	22.68

[a]In 1977, all group members including the three infants were observed for 1 hr each (except female FS, who had been removed).

were assumed to be the older of the pairs of peers. In addition, AB's brow "diadem" was orange like FB's while AS's and FS's were maroon; AS and FS also bore a striking facial similarity. Thus, subadult female AB and juvenile males B1 and C1 were considered to be the offspring of female FB while AS, B2, and C2 were considered to be FS's offspring. These presumed maternities were consistent with proximity patterns identified subsequently.

In spite of the large size of the group and the relatively small enclosure, overt aggression was minimal. However, after the completion of the main portion of the present study, adult female FS received a serious bite wound on her hind leg and had to be removed in April 1977. AS also received a wound on her hind leg but was not removed. Attempts to reintroduce FS failed as she was harassed by group members, especially by FB, AB, and the Bs and Cs. The group was taken off display in 1981 after several individuals died of a rare virus. A DeBrazza's group partially composed of the survivors was put back on display in 1983.

Housing and Maintenance

The group was housed in an indoor enclosure of the Primate House [see Fig. 2 in Oswald and Lockard, 1980]. The cage was approximately 6.28 × 3.96 × 3.5-m high, constructed of concrete and wire mesh with skylights above. A 1.84 × 1.84 × .78-m high "hideout" was accessible from the north end of the cage, and the monkeys were enclosed there during maintenance of the central enclosure. The floor of the cage was covered with straw, which was changed weekly. A large spiral metal perch with round platforms attached at two heights occupied the center of the northern half of the enclosure. Other cage "furnishings" included a vertically rotating "ferris wheel" at the south end, a 1-m-wide ledge 1.9 m from floor level along the back (west) wall, several horizontal bars 1.9 m from the floor, and various branches and logs whose locations were changed periodically. Twice daily, mid- to late morning and mid-afternoon, the monkeys were fed a diet of fresh fruits and vegetables, sunflower seeds, hard-boiled eggs, and monkey chow.

Data Collection Procedure

The principle subjects were three infants (DS, EB, and ES) observed one at a time in random order using a focal animal sampling technique [Altmann, 1974; Crockett, 1978]. All onsets of behaviors directed toward, and initiated by, the focal infant were recorded on a checksheet.

Initially infants DS and EB were each observed for 2.67 hr per week of 20-min focal sample periods (eight per week) (Table II). Subsequent observations included a newly-born infant ES, and each infant was observed an average of 1 hr per week using 10-min focal sample periods (six per week). After adult female FS's removal, each group member was observed in random order for 5-min focal sample periods each week, for a total of 1 hr per monkey. Observation periods alternated morning (0900–1100 hr) and afternoon (1300–1800 hr), and usually occurred 3 times per week.

At the beginning and end of each observation period, the locations of all individuals were recorded on a map of the enclosure. From this scan sample mapping [Altmann, 1974; Crockett, 1978], interindividual distances for all possible dyads were estimated. Distance data are based on 99 maps from 1976 and 17 maps from 1977. Additional proximity data were provided by three unpublished papers by University of Washington students who observed the same group (Taylor, 14 hr in autumn, 1974; Robinson, 11.25 hr in summer, 1975; Rhodes, 16.25 hr in autumn, 1975–winter, 1976).

Data Analysis and Presentation

In Tables III–IV, data for some of the age/sex classes have been combined into sums. The average rate per individual can be estimated by dividing each of the sums by the number of monkeys per class. These tables arrange sample periods on infants in order of their ages during observations. Thus, the 7–10-week sample on EB and 43–46-week sample on DS occurred prior to ES's birth (Table II). Rates (frequency per hr of observation) were calculated separately for infant as actor and recipient.

Interindividual distance and enclosure use (location) data came from the maps. The percent of time that the infant was in contact with the mother was estimated by the percent of maps showing zero distance between mother and infant. The monkeys' locations plotted on the maps were scored as "floor" (including logs), "ledge," "bar" (including ferris wheel and perch), "branch," and "platform."

Interindividual distance data were analyzed using SPSS [Statistical Package for the Social Sciences; Nie et al, 1975]. Each "case" consisted of the data from one map, i.e., the distances between each of 66 possible dyads. Based on mean interindividual distances, "nearest" and "next nearest" neighbor distances were compared with distances to the remaining nine individuals using a paired t-test statistic [Nie et al, 1975]. This test was selected because at a particular point in time, the distances between pairs of animals in the same group are not independent. For example, when the mean distance between MD and FB was compared to the mean distance between MD and EB, only data from those maps that included all three individuals

TABLE III. Hourly Rates of Affiliative Body Contact *Toward* Infant and Rejection *by* Infant (Top Half); Affiliative Body Contact *by* Infant and Rejection *toward* Infant (Bottom Half)[a]

		Affiliative body contact				Rejection			
Subject (sex)	Age (weeks)	Mom n = 1	Other fem. n = 1	SA fem. n = 2	All other n = 6–7[b]	Mom n = 1	Other fem. n = 1	SA fem. n = 2	All other n = 6–7[b]
ES male	1–4	8.7	7.5	12.0	3.9	0	.9	4.5	0
ES male	5–8	2.3	2.5	11.1	7.1	0	.3	2.0	0
EB fem.	7–10	10.9	5.2	7.2	2.6	0	3.5	1.0	.3
ES male	9–12	2.6	1.8	3.1	2.7	0	.8	0	0
EB fem.	14–25	2.8	1.1	1.3	2.7	0	.5	.2	0
ES male	33–46	—	0	1.0	3.0	—	0	0	0
DS male	43–46	1.2	0	.6	2.0	0	0	0	.3
EB fem.	47–60	0	—	0	2.2	0	—	0	0
DS male	50–61	.1	.2	.2	4.1	0	0	0	0
DS male	83–96	–	0	.5	.5	—	0	0	0
ES male	1–4	3.6	.9	3.0	5.7	.6	0	0	0
ES male	5–8	9.0	1.0	8.8	5.6	1.0	0	0	.3
EB fem.	7–10	12.4	1.6	4.6	4.0	0	.1	.2	.3
ES male	9–12	6.0	2.0	3.5	4.4	1.8	0	.5	1.1
EB fem.	14–25	9.1	1.4	3.2	5.3	.1	.3	.1	.7
ES male	33–46	—	0	2.5	5.5	—	.5	0	.5
DS male	43–46	4.1	.2	.2	4.0	2.2	0	.1	.4
EB fem.	47–60	.5	—	.5	2.0	.5	—	0	0
DS male	50–61	1.5	.8	4.4	9.7	.2	0	.1	.6
DS male	83–96	—	0	.5	2.5	–	0	0	0

[a]SA, subadult. Average rate of interaction per dyad can be calculated by dividing these rates by n, the number of monkeys in each category of interactees.

[b]ES had not yet been born when EB was 7–10 weeks old and DS was 43–46 weeks old; "—" indicates period after mother of ES and DS was removed.

were used, and the distance between MD and EB was paired with the distance between MD and FB from the same map.

RESULTS

Physical Appearance of Age/Sex Classes

The most striking characteristics of adult *C. neglectus* are the reddish diadem on the forehead and the white beard. The diadem is broad in the middle of the forehead and tapers to the sides of the head. Below the diadem is a thin brow band of black hairs and above it is a thin white stripe and a sharply defined black coronal band, narrowest in the middle and broadening laterally. The adult's dorsal surface is a greyish olive speckled with black.

TABLE IV. Hourly Rates of Grooming *by* (Top Half) and *Toward* (Bottom Half) Infant[a]

Subject (sex)	Age (weeks)	Highest rate w/	Mom n = 1	Other fem. n = 1	SA fem. n = 2	Ad. male n = 1	Juv. males n = 4	Inf. n = 1–2[b]	Total n = 10–11
ES male	1–4	DS	0	0	0	0	0	.3	.3
ES male	5–8	—	0	0	0	0	0	0	0
EB fem.	7–10	Mom	.7	.1	.3	.1	0	0[b]	1.2
ES male	9–12	Mom,FB	.3	.3	0	0	0	0	.6
EB fem.	14–25	Mom	.4	0	.2	0	.5	0	1.1
ES male	33–46	—	—	0	0	0	0	0	0
DS male	43–46	Mom,FB	.2	.2	.1	0	.5	.3[b]	1.3
EB fem.	47–60	Mom	.5	—	0	0	0	0	.5
DS male	50–61	FB	.2	.3	.3	0	.8	.2	1.8
DS male	83–96	EB	—	0	.5	0	0	1.0	1.5
ES male	1–4	AB/AS	1.5	.6	3.3	0	.3	0	5.7
ES male	5–8	EB	.3	1.0	1.0	0	0	1.3	3.6
EB fem.	7–10	Mom,FS	.5	.5	.5	0	0	.3[b]	1.8
ES male	9–12	AB/AS	.3	.3	.8	0	0	0	1.4
EB fem.	14–25	FS	.3	.5	.3	0	.3	.1	1.5
ES male	33–46	AB/AS	—	0	.5	0	.5	0	1.0
DS male	43–46	Mom	1.1	0	0.4	0	.5	0[b]	2.0
EB fem.	47–60	Mom	1.5	—	.5	0	0	1.0	3.0
DS male	50–61	FB	.2	.3	.3	0	.8	.2	1.8
DS male	83–96	MD	—	0	1.5	1.0	.5	0	3.0

[a]SA, subadult. Average rate of interaction per dyad can be calculated by dividing these rates by n, the number of monkeys in each category of interactees.

[b]ES had not yet been born when EB was 7–10 weeks old and DS was 43–46 weeks old; "—" indicates period after mother of ES and DS was removed.

The tail, limbs, hands, and feet are black while the muzzle, beard, and ventrum are white.

The adult pelage takes about 4 years to develop. The pelage of young infants was blond, darker on the dorsal surface than on the ventrum. Their faces, hands, and feet were pinkish, and there was no indication of the brow pattern. Infant male ES's scrotum could be seen clearly until age 2–3 months after which it could no longer be detected.

The yearling male DS's scrotum could not be discerned by an observer standing outside of the cage. His nose was still pinkish in color, and there was no indication of the brow pattern. The facial hairs, top of the head, and dorsal surface were brown; the tail was lighter than the back. The ventrum was lighter than the dorsum, but was not white.

The scrotum of neither 2-year-old male (C1, C2) was evident. The frontal diadem had not yet differentiated from the brow and head hairs. The overall

TABLE V. Hourly Rates of Social Play *by* (Top Half) and *Toward* (Bottom Half) Infant[a]

Subject (sex)	Age (weeks)	Highest rate w/	Mom n = 1	Other fem. n = 1	SA fem. n = 2	Ad. male n = 1	Juv. males n = 4	Inf. n = 1–2[b]	Total n = 10–11
ES male	1–4	Mom	4.8	.3	3.0	0	.3	.6	9.0
ES male	5–8	EB	3.8	0	7.0	0	5.8	10.0	26.6
EB fem.	7–10	DS	6.2	1.8	6.6	.5	7.4	6.8[b]	29.3
ES male	9–12	EB	4.3	2.5	6.5	0.8	19.8	27.3	61.2
EB fem.	14–25	ES	2.0	0.6	3.5	0.3	12.5	14.7	33.6
ES male	33–46	EB	—	1.0	0	0	12.5	18.0	31.5
DS male	43–46	EB	.3	.7	1.8	1.0	20.2	5.7[b]	29.7
EB fem.	47–60	ES	.5	—	.5	1.5	18.5	19.0	40.0
DS male	50–61	C1/C2	.1	1.3	3.0	0.5	21.2	14.0	40.1
DS male	83–96	C1	—	0	.5	5	6.0	1.5	8.5
ES male	1–4	EB	0	0	.3	0	0	.9	1.2
ES male	5–8	EB	1.3	0	1.5	0	3.3	6.8	12.9
EB fem.	7–10	DS	.7	.2	1.8	.1	2.3	4.2[b]	9.3
ES male	9–12	EB	.8	.3	3.5	0	9.8	21.2	35.6
EB fem.	14–25	DS	.2	0	.4	.2	7.4	11.8	20.0
ES male	33–46	EB	—	0	0	0	14.5	19.0	33.5
DS male	43–46	C1/C2	.1	0	.7	.7	15.3	5.6[b]	22.4
EB fem.	47–60	ES	.5	—	.5	.5	15.0	18.5	35.0
DS male	50–61	EB	0	0	.2	.4	15.2	16.9	32.7
DS male	83–96	C1	—	0	0	0	8.5	1.0	9.5

[a]SA, subadult. Average rate of interaction per dyad can be calculated by dividing these rates by n, the number of monkeys in each category of interactees.

[b]ES had not yet been born when EB was 7–10 weeks old and DS was 43–46 weeks old; "—" indicates period after mother of ES and DS was removed.

coat color was brown with a white ventrum. The white beard first became distinct at about 2 years of age.

The testicles and blue scrotum of the 3-year-old males (B1, B2) were visible. The frontal diadem was beginning to differentiate from the brow and head hairs, but the black brow and coronal band had not yet appeared. The overall coat color was more greyish-olive than brown. The 3-year-old males were smaller in body size than the 4-year-old females.

The 4-year-old females (AB, AS) had acquired adult pelages but were of a more slender build than the adult females. The adult females were considerably smaller than the adult male. Adult male DeBrazza's are reported to weigh 7 kg compared to 4 kg for adult females [Gautier-Hion, 1975].

Reproductive History of Females

All births recorded for females FB and FS between 1967 and 1980 occurred March through November (Table I). Interbirth intervals between successive births for the same female averaged 13.1 months (SD = 1.1; range 11.4–15.6 months, n = 15; excluding interval after 1975 stillbirth). Both females gave birth about a year after their acquisitions, and annually thereafter. All but three of the 18 infants with complete records survived at least through 1 year of age (17% infant mortality). Female FB was born prior to 1962 and was at least 19 years of age when she last gave birth.

The subadult females were not observed engaging in sexual behavior during this study, and they were 5 years of age at its termination. However, AS displayed sexual behavior in 1978 at about 6 years of age.

Infant Development

Contact with mother. Figure 1 presents contact data from this study and previous observations of the zoo group. For the zoo's DeBrazza's, time in contact with mother declined steadily from the first weeks of life. By 12 weeks of age, all four male infants were in contact less than 25% of the time. FB's female infant EB was in contact more often than similar-age male infants for most periods. Contact with the mother remained at 10–25% until the infants were nearly a year of age. DS's contact dropped to zero at age 45 weeks, about 4 weeks before his male sibling ES was born. CS was in contact with FS during about 25% of samples when he was 45 weeks old, about 6 weeks before his male sibling DS was born.

Nursing and foraging. As would be expected during the first 2 weeks of life, the most frequent behavior of the DeBrazza's infant was nursing. As early as the third week of life, nursing became less important. By the fifth week, nursing dropped to below 12% of behavior onsets, and foraging began. By 24 weeks of age, foraging was consistently more frequent than nursing. The yearling DS still nursed occasionally at age 43–45 weeks but rarely nursed after the birth of his male sibling ES. DS's last recorded nursing occurred when he was 57 weeks old. Female EB was not observed to nurse during the 1977 sampling periods when she was 47 to 60 weeks of age. One nursing attempt was rejected by FB when EB was 50 weeks old, about 2 months prior to the birth of FB's next infant. Male ES was abruptly weaned at age 7 months when his mother was removed.

Affiliative body contact and rejection of contact attempts. Affiliative body contact (touching, muzzling, huddling with, picking up, and holding) and rejection (moving from or pushing away, sometimes accompanied by infant's screaming) primarily involved interactions between the two younger infants and the adult and subadult females (Table III). The adult and subadult females frequently initiated contact during infant ES's first month of life. Many of the subadult A's and female FB's early attempts at body contact and

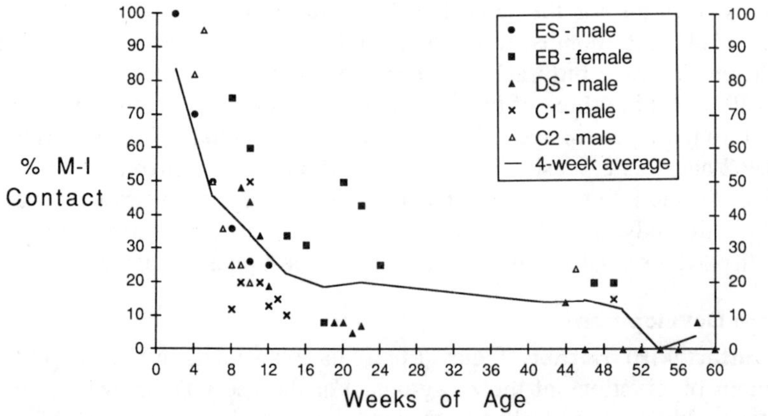

Fig. 1. Mother-infant contact by weeks of age. Percent of time that infant DeBrazza's were in contact with the mother estimated by percent of maps where mother-infant distance was zero (ES, EB, and DS age 43–61 weeks). Other data points from previous studies [Taylor, 1974; Robinson, 1975; Rhodes, 1976]. Line presents 4-week averages (plotted at midpoint of 4-week block), calculated such that each infant observed contributed equally.

holding were rejected by infant ES who squirmed, moved away, and/or screamed. However, his mother FS permitted them to hold ES after they lip-smacked and tugged on the infant. During ES's first month, one of the subadult females sometimes succeeded in holding him for extended periods, exceeding the 10-min observation period. All 231 recorded instances of individuals carrying the three infants were ventral carries. Most were by the infant's mother.

ES's early attempts at affiliative contact with others were rarely rejected. Only his mother was observed to reject him during his first month. Maternal rejection steadily increased, suggestive of the weaning process.

Female infant EB had a high rate of affiliative body contacts with her mother FB, who seldom rejected her (Table III). FS frequently initiated body contact and picked up and held EB before her own infant ES was born; EB rejected many of these attempts. The As also interacted with EB during this time. All adult and subadult females reduced their interactions with EB after ES was born, when EB was 14 weeks old.

Yearling male DS frequently initiated contact toward his mother during the weeks prior to his brother ES's birth, but she rejected him at a high rate. During this time, she initiated more contact with EB than toward her own son. After ES's birth, DS's contacts were more frequent with the other immatures and EB.

Grooming. Infant male ES was first observed to groom another monkey (his brother DS) at age 3 weeks. However, he subsequently groomed just the two adult females, and only once each, during sample periods (Table IV). ES was groomed about as often by FB as by his mother FS. Female EB also groomed infrequently; she groomed her mother most and was groomed most by the adult and subadult females. Infant DS also seldom groomed. His mother groomed him frequently prior to ES's birth, but rarely thereafter.

Grooming data collected on the entire group in 1977 indicated that social grooming was initiated on average 1.9 times per hr per monkey (range 0 by ES to 7.0 by AS). Thus, the infants groomed less than average, and ES was groomed at more than twice the average rate through 8 weeks of age (Table IV).

Social play. Table V presents the hourly rates of social play directed by, and toward, the infants. Social play included playface, play chase, wrestling, grasping and poking, gnaw wrestling, pulling or shaking limbs, horizontal head shake, gnawing, and biting. Each time contact was broken and another play behavior initiated, another score was recorded. Hence, these rates represent play events rather than play bouts.

All three infants played primarily with each other and with the juveniles (Table V). The adult male rarely initiated play and never did so toward the youngest infant ES. When ES approached MD, the adult male either gently pushed the infant away or moved away from him. ES and EB played more with their own mothers than with the other adult female, and these interactions were usually initiated by the infants. The subadult females were the recipients of an appreciable amount of social play from ES and EB when these infants were under 6 months of age, but DS played with the subadults less often. EB played most with the C monkeys and DS until ES was about 3 months old, and then he became her most frequent partner.

DS's rate of play declined in 1977 when he was 83–96 weeks old. He initiated play events only 8.5 times per hr compared to 22.0 events per hr by C2, 18.0 by C1, 11.0 by B2, and 4.0 per hour by B1. The Cs played more with each other and with the younger infants EB and ES than with DS. DS's lower rate of play may have been partially due to the fact that he had no play partner his own age because of FB's 1975 stillbirth.

The frequency of play dropped sharply with age. The adult male and the adult and subadult females initiated play only .5 (MD) to 2.0 (AS) times per hr.

Agonistic Interactions

Physical aggression was rare. The only agonism recorded during sampling periods took the form of threats and submissive responses. Threats included open mouth stare with and without accentuation of brow; lunging with tail raised and head lowered and arms held stiffly (sometimes accompanied by hopping); chasing; and bipedal stance. Submission included conspicuous eye aversion, lip-smacking, and grimace; being supplanted accompanied by one of the preceding facial expressions; presenting or crouching.

Infants' agonistic interactions. ES's first recorded submissive response occurred when he was 9 weeks old (he showed a submissive crouch), and he received his first recorded threat at age 12 weeks (this was in the form of a threat stance). DS performed the most submissive responses recorded for the three infants in 1976, a total of 31 for an hourly rate of 1.4; 12 were toward the adult male MD and nine were toward the C monkeys; the rest were toward the As (4), the Bs (3), FB (2), and EB (1). EB directed 16 submissive gestures toward others in 1976 (.7 per hr): toward MD (5), FS (4), the Bs (3), the Cs (2), and once each to her mother FB and an A. The youngest infant ES showed submission once to a C and once to EB (.2 per hour).

Dominance hierarchy of group. Table VI presents threats and submission recorded for all group members during 1977 (after adult female FS's removal). Submissive gestures were nearly twice as frequent as overt threats. The overall rate was 3.18 agonistic acts per hour per monkey. The adult male MD gave 38.5% of all threats and received 30.4% of all submission; he never was threatened by others. Subadult female AS performed the most submissive behaviors, 23.9%, and received 15.4% of all threats. Infant EB was not involved in any agonistic interactions. In Table VI, the monkeys (except EB) are ordered based on the ratio of interactions when they were dominant (delivered threat plus received submission) to total agonistic interactions in which each was involved. This yields a fairly linear hierarchy in which all of FB's offspring ranked above FS's except for AB, and she was dominant to her peer AS.

Although the group data were collected after FS's removal, FB was clearly dominant to FS. FB never was observed to be threatened or chased by any individual with the exception of adult male MD, and then only once. In contrast, FS was regularly threatened by MD and AB during feeding times. On a few occasions she was chased around the cage by AB and one of the B monkeys. These chases probably culminated in the wound which led to FS's eventual removal.

TABLE VI. Agonistic Interactions, 1977 Data[a]

Dominant: Delivered threat or (received submission)	Subordinate: Received threat or (delivered submission) MD	FB	C1	B1	C2	B2	DS	AB	AS	ES	Total dominant		%[b]
MD	—	1	1	1(2)	1(1)	2(3)	(2)	2(1)	1(1)	1(4)	10(14)	24	100
FB		—	1		(1)	1(1)					2 (2)	4	80
C1			—	3	(1)		(1)			(2)	3 (2)	5	71
B1				—		(2)		2(6)	1(2)		3(10)	13	62
C2					—	1(2)	1(1)		(1)	(2)	2 (6)	8	53
B2				1	1(1)	—			1(4)	(1)	3 (6)	9	43
DS				1	1		—		(1)		2 (1)	3	33
AB								—	1(2)		1 (2)	3	21
AS							(1)		—		0 (1)	1	6
ES										—	0 (0)	0	0
Total	0(0)	1(0)	2(0)	6(2)	3(4)	4(8)	1(5)	4(7)	4(11)	1(9)	26(46)		
Subordinate	0	1	2	8	7	12	6	11	15	10	72		
Percent[c]	0	20	29	38	47	57	67	79	94	100			

[a]Each cell based on 2-hr observation (each monkey was focal subject for 1 hr). FS was removed prior to 1977 observations. EB was not involved in any recorded agonistic interactions.

[b](Delivered threat + received submission) ÷ total agonistic interactions for individual.

[c](Delivered submission + received threat) ÷ total agonistic interactions for individual.

Enclosure Use

Individuals differed in their use of various locations within the enclosure (Fig. 2). The adult male used the branches most often and the floor least compared to the other monkeys. He was observed in arboreal (nonfloor) locations significantly more often than female FB ($\chi^2 = 9.99$, d.f. $= 1$, $p < .002$) [Siegel, 1956], and FB was herself arboreal significantly more often than female FS ($\chi^2 = 8.37$, d.f. $= 1$, $p < .01$). The members of the subordinate FS matriline were on the floor of the enclosure 62% of the time compared to 45% for FB and her offspring ($\chi^2 = 17.95$, d.f. $= 1$, $p < 0.001$).

Interindividual Distance and Proximity Patterns

The maximum separation possible in the zoo enclosure was about 7 m. Of 66 total mean interindividual distances calculated, the smallest average was 1.28 m between FS and her youngest infant ES. The largest mean distance was 3.35 m between 3-year-old male B1 and 2-year-old male C2. There were

DISCUSSION

Infant Development

In captivity, DeBrazza's monkey infants begin interacting with individuals other than their mothers very early in life. Most of these early interactions involve affiliative contact and social play. Foraging begins during the second month of life and is consistently more frequent than nursing by the fifth month. By 1 year of age, the infant is largely independent and may infrequently interact with its mother if a younger sibling has been born.

The picture of DeBrazza's development from the present study is generally compatible with data from Chalmers [1972] and Stevenson [1973]. Our zoo DeBrazza's infants showed a rapid decline in the percent of time spent in contact with their mothers (Fig. 1). The data presented by Chalmers [1972] show a similar decline in contact, but his percentages are considerably lower for the first 7 weeks. This is undoubtedly because his figures represent the percent of time in which the mother *held* the infant while ours include all types of contact. The present study provides no data on emergence of locomotor skills since the three infants were not observed from birth. The other studies indicate that DeBrazza's infants generally first leave their mothers during the first week of life, at 2, 4, 8 [Chalmers, 1972], 5, and 7 days of age [Stevenson, 1973]. Infant ES was first observed to groom another animal at age 3 weeks whereas the infants in Stevenson's [1973] zoo study were about 10 weeks old. Infant ES also initiated social play and foraging younger than Stevenson's [1973] two infants. In both zoo studies, infants survived premature weaning caused by the removal (this study, at 7 months of age) or death [Stevenson, 1973, at 5 months of age] of their mothers. Normal weaning of DeBrazza's infants appears to occur at about 1 year of age.

Comparisons With Other Cercopithecus Species

The general course of development found for DeBrazza's infants is similar to that reported for wild *Cercopithecus campbelli lowei* [Bourliere et al, 1970] and *C. aethiops* [Struhsaker, 1971]. Infant *C. campbelli* can forage independently when 3 to 4 months old. They are weaned gradually about 2 months before the next infant is born, usually at about 10 months of age; however, infants have been observed to suckle until 26 months of age if there is no younger sibling. The yearlings are extremely agile and only occasionally are carried ventrally by the mother [Bourliere et al, 1970]. Infant vervets' nursing attempts are terminated by mothers' nips most frequently when they are 3 months old [Struhsaker, 1971], although infants up to 266 days (ca. 9 months) have been observed suckling [Struhsaker, 1967].

Thus, vervets appear to be weaned somewhat younger than *C. campbelli* and *C. neglectus*.

The rate at which captive DeBrazza's mothers groom their infants (.2–1.5 per hr) is similar to the rate that wild vervet mothers groom their infants [.18–3.1 grooms per hr, Struhsaker, 1971]. Chalmers [1972] found that captive *C. aethiops*, *C. neglectus*, and *C. mitis albogularis* (Sykes') infants first displayed selected motor activities at similar ages.

Play was the most frequent behavior initiated by the infants in the present study. Play in a wild DeBrazza's troop was not observed during 320 hours of observation, but the troop included only one infant [Gautier-Hion and Gautier, 1978]. Mörike [1976] and Stevenson [1973] describe play behaviors of captive *C. neglectus* in some detail, and they are similar to our own observations as well as to behaviors displayed by other cercopithecines. Play in the zoo DeBrazza's declined with age but remained frequent through 2 years of age. Hourly rates of play in this study are much higher than those reported for similarly aged vervets and Sykes' guenons by Bramblett and Coelho [this volume], but their figures are for play *bouts* while ours are for play *events* (see Methods). Fady [1969] found that the offspring of high ranking *Macaca fascicularis* females played more often than the offspring of low-ranking females. There was no clear relationship between rank and the amount of play among the DeBrazza's.

Infant Independence: Theories and Evidence

Infants show intraspecific and interspecific variation in the rate at which they attain independence from their mothers, often measured by the percent of time that infants are in contact with their mothers at different ages. Three general hypotheses have been proposed: 1) the arboreal versus terrestrial theory; 2) the physical maturation–age at puberty theory; and 3) the maternal restrictiveness theory. The first theory predicts that infants of arboreal species should become independent of the mother more slowly than those of terrestrial or semiterrestrial species because uncoordinated infants falling from tall trees are more likely to be injured than those falling short distances [Chalmers, 1972]. The second theory predicts that infant independence tracks maturation, and whatever factors favor younger ages at puberty ought to favor more rapid infant independence. The third theory predicts that infants with more restrictive mothers will become independent more slowly than those with more permissive mothers; this theory could potentially account for both intraspecific and interspecific variation in infant independence while the others only address interspecific variation.

Comparable data estimating percent of time that infants of different ages are in contact with their mothers are available for six cercopithecine species: captive [Fairbanks, unpublished] and wild [Struhsaker, 1971] *C. aethiops*; captive *C. neglectus* (Fig. 1, this study); captive patas monkeys [*Erythrocebus patas*; Loy and Loy, this volume]; wild yellow baboons [*Papio cynocephalus*; Altmann, 1980]; free-ranging (island) rhesus [*Macaca mulatta*; Berman, 1978, cited in Altmann, 1980]; and a group of captive pigtailed macaques [*M. nemestrina*, Castell and (Crockett) Wilson, 1971]. The baboon, patas, and pigtail infants develop independence most slowly, and in a very similar pattern: During the first month of life, they spend more than 90% of the time in contact with their mothers; at age 3 months, they are in contact about 60% of the time; by 5–6 months, contact has dropped to about 30%. Vervet infants spend less time in contact at comparable ages, and wild and captive vervets are very similar: During the first month, vervets are in contact about 80% of the time, dropping to about 30% at 3 months and about 15% by 5–6 months. Rhesus infants' courses of independence are similar to vervets during the first month (ca. 80% contact) but decline to only 40% by 3 months and resemble the baboons, patas, and pigtails by 5–6 months when they are still in contact with their mothers about 30% of the time. The DeBrazza's infants in the present study resemble the vervets during the first month, but their percent of time in contact with the mother declines more sharply to only 23% of the time at age 3 months; at 5 months, however, their contact percentage is about the same as vervets. Table VII presents Spearman's correlations between the rank order of mother-infant contact at age 3 months for these six species with rank orders of other variables pertinent to the infant independence theories.

The arboreal-terrestrial theory. These contact data seem to rule out the arboreal-terrestrial theory. Chalmers' [1972] data appeared to support this theory in that the more terrestrial vervets and DeBrazza's infants showed more rapid independence than the predominantly arboreal Sykes' (*C. m. albogularis*) and mangabey (*Cercocebus albigina*). (Unfortunately, his data are for percent of time held by mother, not all contact combined, so they cannot be directly compared with the above figures.) However, of the six species with comparable contact data, the baboons and patas are the most terrestrial while vervets and rhesus are intermediate and semiterrestrial, spending a good deal of the day on the ground; DeBrazza's are the most arboreal of the six species and probably spend somewhat less time on or near the ground than do pigtail macaques, who themselves appear to be terrestrial less often than most rhesus [Altmann and Altmann, 1970; Gartlan, 1974; Dunbar and Dunbar, 1974; Kavanagh, 1978; Neville, 1968; Quris, 1976; Gautier-Hion and Gautier, 1978; Crockett and Wilson, 1980]. Thus, the

TABLE VII. Testing Hypotheses of Infant Independence[a]

Species	Independence[b]	Terrestriality[c]	Age at 1st birth[d]	Adult weights[e] Female	Adult weights[e] Male	Infant handling[f]
C. aethiops	2	3.5	2	2	1	3
C. neglectus	1	6	5	3	3	1
E. patas	4	1.5	1	4	4	3
M. mulatta	3	3.5	3	1	2	3
M. nemestrina	6	5	4	5	5	5.5
P. cynocephalus	5	1.5	6	6	6	5.5
r_S with Independence		−.34	.14	.71	.77	.93
Probability		(N.S.)	(N.S.)	0.11	0.08	0.036

[a]Variables assigned ranks based on quantitative data or qualitative assessment as indicated. Spearman rank order correlation coefficients (r_S) and probability levels (n = 6 for all comparisons).

[b]Most to least independent at 3 months of age; based on percentage of mother-infant contact from sources cited in text.

[c]Most to least terrestrial; estimated rank order of percent of time spent on the ground, from sources cited in text. Tied ranks presented when information insufficient to rank one species above another.

[d]Youngest to oldest at first birth; based on ages given in sources cited in text.

[e]Lightest to heaviest; based on weights given in Harvey and Clutton-Brock [1985]. For these six species, male and female weights were significantly correlated (r_S = .94, p = 0.033) and both were equally correlated with infant handling (r_S = .70, p = 0.11).

[f]Least to most vulnerable to potentially detrimental handling of infants; based on estimates of potential number of available handlers per vulnerable infant. Tied ranks presented when information insufficient to rank one species above another.

degree of terrestriality is not consistently related to the attainment of infant independence in the predicted direction, even within genera. However, none of these species is highly arboreal, and additional within-genus comparisons might shed further light on this hypothesis. For example, the arboreal mangabey *C. albigena* could be compared with *Cercocebus galeritus*, which Quris [1976] found to be on or near the ground about the same percentage of time as were DeBrazza's. The genus *Macaca* also varies in terrestriality, even among forest species; for example, *M. nemestrina* is more terrestrial than *M. fascicularis* [Crockett and Wilson, 1980], and *M. silenus* is the most arboreal of the macaques [Green and Minkowski, 1977].

The physical maturation–age at puberty theory. Differential attainment of independence by *Macaca* versus *Cercopithecus* species is probably unrelated to early physical maturation and coordination, since macaque infants, like vervets and DeBrazza's, can locomote independently within several days of birth [e.g., 4 days, *M. fuscata*, C. Crockett, personal observation; *M. mulatta*, walking at 2 days, climbing at 4 days, Castell and Sackett, 1973]. Furthermore, rhesus, vervets, and patas have discrete birth seasons, and many females give birth annually [Drickamer, 1974; Struhsaker, 1967; Chism et al, 1984], suggesting that the period of dependency and maximum maternal investment is similar for all three species. The DeBrazza's females in the present study also gave birth annually. Baboons, on the other hand, have interbirth intervals of about 21 months [Altmann et al, 1977], yet show mother-infant contact percentages very similar to patas.

It is possible that the rate of social development and independence is related to age at puberty. There is a significant variation in age at puberty among African cercopithecines [Rowell, 1977; Rowell and Richards, 1979; Chism et al, 1984]. Rowell [1977] suggests that cercopithecines in relatively invariant forest habitats reach sexual maturity later than species adapted to habitats with less predictable or more heterogeneous climates and food supplies. Primate species adapted to successional habitats should show "*r*-selected" characteristics [Pianka, 1970] such as bearing the first offspring at a younger age than "*K*-selected" primary forest adapted species [Rowell and Richards, 1979; Crockett and Wilson, 1980]. Consistent with this idea, savanna and woodland-savanna dwelling patas and vervets first give birth at modal ages of about 3 and 4 years, respectively [Rowell and Richards, 1979; Chism et al, 1984], while forest-dwelling African cercopithecines reach sexual maturity during their fourth year and usually first give birth at about 5 years of age (*Cercopithecus neglectus*, *C. mitis*, *C. cephus*, *C. pogonias*, *C. nictitans*, *Cercocebus albigena*, *C. galeritus*, *C. torquatus*, and *Miopithecus talapoin*) [Rowell, 1977; Rowell and Richards, 1979; Gautier-Hion and Gautier, 1976; Bramblett et al, 1975]. Puberty varies in the same direction among macaques: *Macaca mulatta*, which occupy a wide range of habitat types, average 4.0 years at first birth with 21% giving birth during their third year [Drickamer, 1974]. Evergreen forest macaques appear to be older at first birth, averaging 4.5 years [Kempske, 1985] to 5.25 years (Woodland Park Zoo records) for *M. silenus*, and about 4.8 years for first captive generation *M. nemestrina* [Sackett, personal communication]. (Interpretations of age at first birth are complicated by the finding that, at least among *M. nemestrina*, age at first conception declined significantly from first to second and second to third generation females born in captivity; Sackett, personal communication.) Yellow baboons are savanna dwellers but of much larger body size than the other monkeys; they first give birth at about 6.5 years of age [Altmann et al, 1977]. There are positive correlations between body size and various reproductive characteristics, such as age at weaning, age at first birth, and interbirth interval [Harvey and Clutton-Brock, 1985]. Table VII shows that for the six species examined, adult body size is not a good predictor of female's age at first reproduction, nor is first reproduction a good predictor of infant independence, contrary to the maturation theory. However, adult weights show positive correlations with infant independence at 3 months of age, which approach statistical significance (Table VII). Thus, there is some support for the idea that rate of infant independence is related to physical development, but that adult body size rather than age at sexual maturity is the more important variable.

The maternal restrictiveness theory. Infant independence could be retarded by maternal restrictiveness or accelerated by maternal rejection of infants' contact attempts. Restraint by the mother appears to delay infant independence of captive and wild baboons, captive pigtail macaques, and captive vervets [Rowell et al, 1968; Altmann, 1980; Castell and (Crockett) Wilson, 1971; Fairbanks, unpublished], although it did not appear to do so in the captive vervets, DeBrazza's, Sykes', and mangabeys studied by Chalmers [1972]. High maternal rejection was related to earlier infant independence and shorter interbirth intervals in captive rhesus macaques [Simpson et al, 1981]. In the present study of captive DeBrazza's, neither mother was very restrictive, but the infant that was rejected least (female EB) was in contact with her mother FB more of the time than the infants (males ES and DS) of the more frequently rejecting mother FS (Table III; Fig. 1).

Differential maternal restrictiveness and/or rejection does account for some of the intraspecific variation in attainment of infant independence. However, are species differences in restraint and rejection consistent with *interspecific* variation in independence? One possibility is that restrictiveness ought to be greater by individual mothers and species for which handling of infants by other group members is more likely to be detrimental. Baboons and vervets provide intraspecific evidence for this idea in that high-ranking females (who ought to be better able to terminate unwanted handling of their infants) are less restrictive than low ranking ones [Altmann, 1980; Fairbanks, unpublished]. Similarly, group-living pigtail infants were not seen to leave their mothers until they were 1 month old, while infants caged alone with their mothers left them much earlier [Castell and (Crockett) Wilson, 1971]. "Allomaternal" handling of vervets and patas infants is common [Kavanagh, 1983; Zucker and Kaplan, 1981]. The amount of potentially detrimental [e.g., see Wasser and Barash, 1981] handling that an individual infant receives may be reduced by synchronized breeding in vervets and patas, and perhaps rhesus, since handling would be divided among more infants [Kavanagh, 1983; Chism et al, 1984; Drickamer, 1974]. Baboons and pigtails are not strongly seasonal, and births within the same troop do not appear to be highly synchronized [Altmann, 1980; Crockett and Wilson, 1980]. The Woodland Park Zoo DeBrazza's did not show pronounced seasonality, but the two females tended to synchronize their births (Table I). In the wild, however, troops infrequently have more than one adult female [Gautier-Hion and Gautier, 1978; Brennan, 1985]. Thus, if we assume that infant handling is detrimental in proportion to its potential occurrence, DeBrazza's should be least vulnerable, baboons and pigtails should be most vulnerable, while patas, vervets, and rhesus should be intermediate. Development of infant independence is slowest in the species most vulnerable to harm by allomothers and fastest in the least vulnerable DeBrazza's, but the intermediate species show a range of independence patterns (Table VII). Overall, the infant handling variable shows the only significant correlation with infant independence. However, handling is also somewhat correlated with adult weight (see Table VII footnote).

In summary, no single theory is unequivocally supported. For data available on these six cercopithecine species, potential degree of infant handling and adult weight are the best predictors of the patterns of infant independence, while degree of terrestriality and age at puberty seem unrelated.

The Behavior of the Group

Observations of the behavior of the DeBrazza's monkey group at Woodland Park Zoo are similar to those reported by Chalmers [1973] and Mörike [1976]. All three studies indicate virtually no aggression toward the adult male. The rate of aggression was lowest in Chalmer's [1973] captive group (about 0.1 acts/hr/monkey). The rate of agonism in the WPZ group in 1977 was 3.18/hr/monkey, somewhat higher than that reported by Mörike [1976] (about 1.5/hr/monkey). In the Woodland Park Zoo group, the adult male delivered the most threats, while in the Heidelberg Zoo group, the adult female was most aggressive [Mörike, 1976]. No aggression was observed in wild DeBrazza's [Gautier-Hion and Gautier, 1978].

In 1976, serious aggression was very infrequent in the WPZ DeBrazza's monkeys. Robinson [1975] also reported low aggression levels for this group. Ad libitum [Altmann, 1974] observations during 1976 indicated that the majority of the threats were directed toward the subordinate adult female FS and her subadult daughter AS. This aggression became more physical, and FS was eventually removed due to injuries. Quantitative data in 1977 showed that all of FS's offspring ranked below their peers. It could be argued that this was because FS was no longer present to defend them. However, FS and AS had both been injured by attacks from other group members prior to FS's removal, providing evidence to the contrary.

Although the zoo setting is artificial, we suspect that the gradually increasing aggression directed rather selectively toward the subordinate matriline may reflect patterns found in the wild. DeBrazza's in their natural habitat live in very small troops, about half the size of the zoo group, and often contain only one adult female [Gautier-Hion and Gautier, 1978; Brennan, 1985]. With the exception of *C. l'hoesti preussi*, which appear to live in similar sized troops, the remaining forest species of *Cercopithecus* live in troops of seven to 20 [Struhsaker, 1969] or larger [*C. mitis stuhlmanni*, 13–27, Rudran, 1978; *C. ascanius schmidti*, about 35, Struhsaker, 1977]. If the tendency for DeBrazza's to form small troops is innate, then one would expect aggression to result when group size becomes too large. In the wild, such aggression might lead to emigration of subordinates; in captivity, the end result may be the same either through death or removal of those persecuted. Furthermore, kin selection theory predicts that such aggression would not be random and would be directed toward unrelated or more distantly related individuals [Kurland, 1977].

Reproductive competition seems implicated in the current observations that much aggression was directed toward the present and future reproductive females of the subordinate matriline. In the Neotropical red howler monkey (*Alouatta seniculus*), many maturing females emigrate from their natal troops [Crockett, 1984]. This is apparently a consequence of female-female competition for a limited number of reproductive positions consistent with an optimal troop size. Crockett [1984] suggests that in red howlers "optimal troop size" is a function of foraging strategy. However, in DeBrazza's monkeys, what limits troop size may be related to this species' antipredator strategy of stealth and freezing [Gautier-Hion and Gautier, 1978]. It is unlikely that *C. neglectus* is obligately monogamous throughout its distribution, since some troops do have more than one adult female [Brennan, 1985]. Nevertheless, unless mortality of adult or maturing females is extremely high, some females must emigrate in order to account for the low number of adult females per troop. Thus, new troop formation of a sort reported for red howlers also may be occurring in DeBrazza's monkeys. This possibility raises a further hypothesis regarding the rapid attainment of independence among DeBrazza's infants: individuals who become independent younger may be better prepared for early emigration from their natal troops. Such early emigration by both sexes might be advantageous to relatives remaining in smaller, more concealed troops as well as to the emigrants themselves who might thereby increase their opportunities for finding or forming a new troop in which to breed.

CONCLUSIONS

1. DeBrazza's monkeys show a pattern of infant development that parallels that of other *Cercopithecus* species. For six cercopithecine species with comparable data on mother-infant contact, DeBrazza's showed the most rapid independence, and vervets had a similar pattern. Pigtail macaques, yellow baboons, and patas infants became independent most slowly, and rhesus infants were intermediate.

2. Several hypotheses that attempt to predict variations in attainment of infant independence were evaluated. No single theory was unequivocally supported by data from the six species. Potential degree of infant handling and adult weight showed the highest correlations with infant independence patterns, while degree of terrestriality and age at puberty seemed unrelated.

3. Play is the predominant behavior of captive infant DeBrazza's; age seems to be the main determinant of play partnerships and frequency.

4. By 2 years of age, matriline membership appears to influence proximity patterns and rank.

5. Emerging field observations suggest that the zoo's DeBrazza's lived in a group about twice as large as those in natural habitats. Rates of social interactions recorded in this study are considerably higher than those seen in the wild.

6. Natural selection favoring small group size as part of an antipredator strategy in the wild predicts the occurrence of female reproductive competition, perhaps accounting for some of the aggression seen the the zoo group. In nature, small groups may in part result from early emigration of natal animals of both sexes and this may in turn favor the rapid attainment of independence by DeBrazza's infants.

ACKNOWLEDGMENTS

We wish to thank the staff of the Woodland Park Zoological Gardens for their cooperation and encouragement throughout this study. Special thanks to L. Gledhill and H. Shewman, Primate Keepers, for access to the zoo records. We also thank P. Taylor, C. Robinson, and P. Rhodes for the use of their mother-infant contact data. We thank B. Bedell for photographing the illustrations, R. Brooks for statistical advice, T. Knight for additional references, and V. Bound for translation assistance. Various versions of the manuscript benefitted from comments by D. Cheney, J. Lockard, E. Zucker, and anonymous reviewers. Computer time at the University of Washington Academic Computer Center was provided by the Department of Psychology. This study was initiated while the senior author was a student in the co-author's course "Behavioral Studies of Zoo Animals" (Psychology 419).

REFERENCES

Altmann J (1974): Observational study of behavior. Behaviour 49:227–267.

Altmann J (1980): "Baboon Mothers and Infants." Cambridge: Harvard University Press.

Altmann J, Altmann SA, Hausfater G, McCuskey SA (1977): Life history of yellow baboons: Physical development, reproductive parameters and infant mortality. Primates 18:315–330.

Altmann SA, Altmann J (1970): "Baboon Ecology." Chicago: University of Chicago Press.

Berman CM (1978): Social relationships among free-ranging infant rhesus monkeys. PhD Thesis, Cambridge University.

Bourliere F, Hunkeler C, Bertrand M (1970): Ecology and behavior of Lowe's guenon (*Cercopithecus campbelli lowei*) in the Ivory Coast. In Napier JR, Napier PH (eds): "Old World Monkeys: Evolution, Systematics and Behavior." New York: Academic Press, pp 297–350.

Bramblett CA, Pejaver LD, Drickman DH (1975): Reproduction in captive vervet and Sykes' monkeys. J Mammal 56:940–946.

Brennan EJ (1985): DeBrazza's monkeys (*Cercopithecus neglectus*) in Kenya: Census, distribution, and conservation. Am J Primatol 8:269–277.

Brown LH, Urban EK (1969): DeBrazza's monkey *Cercopithecus neglectus schlegel* in the forests of south-west Ethiopia. E Afr Wildlife J 7:174–175.

Castell R, (Crockett) Wilson C (1971): Influence of spatial environment on development of mother-infant interaction in pigtail monkeys. Behaviour 39:202–211.

Castell R, Sackett G (1973): Motor behaviors of neonatal rhesus monkeys: Measurement techniques and early development. Dev Psychobiol 6:191–202.

Chalmers NR (1972): Comparative aspects of early infant development in some captive Cercopithecines. In Poirier FE (ed): "Primate Socialization." New York: Random House, pp 63–82.

Chalmers NR (1973): Differences in behaviour between some arboreal and terrestrial species of African monkeys. In Michael RP, Crook JH (eds): "Comparative Ecology and Behaviour of Primates." London: Academic Press, pp 69–100.

Chism J, Rowell T, Olson D (1984): Life history patterns of female patas monkeys. In Small MF (ed): "Female Primates." New York: Alan R. Liss, Inc., pp 175–190.

Crockett C (1978): Methods of observational research in the zoo setting. In Crockett C, Hutchins M (eds): "Applied Behavioral Research at the Woodland Park Zoological Gardens." Seattle: Pika Press, pp 51–73.

Crockett CM (1984): Emigration by female red howler monkeys and the case for female competition. In Small MF (ed): "Female Primates." New York: Alan R. Liss, Inc., pp 159–173.

Crockett CM, Wilson WL (1980): The ecological separation of *Macaca nemestrina* and *M. fascicularis* in Sumatra. In Lindburg DG (ed): ''The Macaques: Studies in Ecology, Behavior and Evolution.'' New York: Van Nostrand Reinhold, pp 148–181.

Drickamer LC (1974): A ten-year summary of reproductive data for free-ranging *Macaca mulatta*. Folia Primatol 21:61–80.

Dunbar RIM, Dunbar EP (1974): Ecological relations and niche separation between sympatric terrestrial primates in Ethiopia. Folia Primatol 21:36–60.

Fady JC (1969): Les jeux sociaux: Le compagnon de jeux chez les jeunes. Observations chez *Macaca irus*. Folia Primatol 11:134–143.

Fairbanks LA: Mother-infant contact and the development of independence in vervet monkeys (*Cercopithecus aethiops sabaeus*), unpublished manuscript.

Gartlan JS (1974): Adaptive aspects of social structure in *Erythrocebus patas*. In ''Symposia of the 5th Congress of the International Primatological Society.'' pp 161–171.

Gautier-Hion A (1975): Dimorphisme sexuel et organisation sociale chez les cercopithecines forestiers africains. Mammalia 39:365–374.

Gautier-Hion A (1978): Food niches and coexistence in sympatric primates in Gabon. In ''Recent Advances in Primatology: Behavior.'' New York: Academic Press, Vol. 1, pp 269–286.

Gautier-Hion A, Gautier JP (1976): Croissance, maturité sexuelle et sociale, reproduction chez les cercopithecines forestiers africains. Folia Primatol 26:165–184.

Gautier-Hion A, Gautier JP (1978): Le singe de Brazza: Une strategie originale. Z Tierpsychol 46:84–104.

Gautier JP, Gautier-Hion A (1977): Communication in old world monkeys. In Sebeok TA (ed): ''How Animals Communicate.'' Bloomington: Indiana University Press, pp 890–964.

Green S, Minkowski K (1977): The lion-tailed monkey and its South Indian rain forest habitat. In Prince Rainier III, Bourne GH (eds): ''Primate Conservation.'' New York: Academic Press, pp 290–337.

Harvey PH, Clutton-Brock TH (1985): Life history variation in primates. Evolution 39:559–581.

Hill WCO (1966): ''Primates: Comparative Anatomy and Taxonomy. Cercopithecoidea.'' New York: John Wiley and Sons, Vol. 6.

Kavanagh M (1978): The diet and feeding behaviour of *Cercopithecus aethiops tantalus*. Folia Primatol 30:30–63.

Kavanagh M (1983): Birth seasonality in *Cercopithecus aethiops*: A social advantage from synchrony? In Seth PK (ed): ''Perspectives in Primate Biology.'' New Delhi: Today & Tomorrow's Printers and Publishers, pp 89–98.

Kempske S (1985): *Macaca silenus*: Survey of North American and European zoo practices. In Heltne PG (ed): ''The Lion-Tailed Macaque: Status and Conservation.'' New York: Alan R. Liss, Inc., pp 221–235.

Kurland JA (1977): Kin selection in the Japanese monkey. Contrib Primatol 12:1–145.

Mörike D (1976): Verhalten einer Gruppe von Brazzameerkatzen (*Cercopithecus neglectus*) im Heidelberger Zoo. Primates 17:475–512.

Napier JR, Napier PH (1967): ''A Handbook of Living Primates.'' New York: Academic Press.

Neville MK (1968): Ecology and activity of Himalayan foothill rhesus monkeys (*Macaca mulatta*). Ecology 49:110–123.

Nie NH, Hull CH, Jenkins JG, Steinbrenner K, Bent DH (1975): SPSS Statistical Package for the Social Sciences: Second Edition.'' New York: McGraw-Hill.

Oswald M, Lockard JS (1980): Ethogram of the DeBrazza's guenon (*Cercopithecus neglectus*) in captivity. Appl Anim Ethol 6:285–296.

Pianka ER (1970): On *r*- and *K*-selection. Am Naturalist 104:592–597.

Pocock RI (1907): A monographic revision of the monkeys of the genus *Cercopithecus*. Proc Zool Soc Lon 677–746.

Quris R (1976): Données comparatives sur la socio-écologie de huit especes de Cercopithecidae vivant dans une même zone de forêt primitive periodiquement inondée (Nord-Est du Gabon). La Terre et La Vie 30:193–209.

Rhodes P (1976): Socialization in *Cercopithecus neglectus*. Unpublished Manuscript, Woodland Park Zoo Files.

Robinson C (1975): Primate social behavior: The DeBrazza's. Unpublished Manuscript, Woodland Park Zoo Files.

Rowell TE (1977): Variation in age at puberty in monkeys. Folia Primatol 27:284–296.

Rowell TE, Din NA, Omar A (1968): The social development of baboons in their first three months. J Zool (Lond) 155:461–483.

Rowell TE, Richards SM (1979): Reproductive strategies of some African monkeys. J Mammalogy 60:58–69.

Rudran R (1978): Socioecology of the blue monkeys (*Cercopithecus mitis stuhlmanni*) of the Kibale Forest, Uganda. Smithsonian Contrib Zool 249:1–88.

Schwartz E (1928): Notes on the classification of the African monkeys in the genus *Cercopithecus* Exleben. Ann Nat History 1:649–663.

Siegel S (1956): "Nonparametric Statistics for the Behavioral Sciences." New York: McGraw-Hill.

Simpson MJA, Simpson AE, Hooley J, Zunz M (1981): Infant-related influences on birth intervals in rhesus monkeys. Nature 290:49–51.

Stevenson MF (1973): Observations of maternal behaviour and infant development in the DeBrazza monkey *Cercopithecus neglectus*. Int Zoo Yearbook 13:179–184.

Struhsaker TT (1967): Behavior of vervet monkeys (*Cercopithecus aethiops*). Univ Calif Publ Zool 82:1–74.

Struhsaker TT (1969): Correlates of ecology and social organization among African Cercopithecines. Folia Primatol 11:80–118.

Struhsaker TT (1971): Social behaviour of mother and infant vervet monkeys (*Cercopithecus aethiops*). Anim Behav 19:233–250.

Struhsaker TT (1977): Infanticide and social organization in the redtail monkey (*Cercopithecus ascanius schmidti*) in the Kibale Forest, Uganda. Z Tierpsychol 45:75–84.

Taylor PK (1974): Socialization of *Cercopithecus neglectus* infants. Unpublished Manuscript, Woodland Park Zoo Files.

Wasser SK, Barash DP (1981): The selfish "allomother": A comment on Scollay and DeBold (1980). Ethol Sociobiol 2:91–93.

Wolfheim JH (1983): "Primates of the World: Distribution, Abundance and Conservation." Seattle: University of Washington Press.

Zucker EL, Kaplan JR (1981): Allomaternal behavior in a group of free-ranging patas monkeys. Am J Primatol 1:57–64.

Comparative Behavior of African Monkeys, pages 67–79

4

Development of Social Behavior in Vervet Monkeys, Sykes' Monkeys, and Baboons

Claud A. Bramblett and Anthony M. Coelho, Jr.

Department of Anthropology, The University of Texas at Austin, Austin, Texas 78712 (C.A.B.), and Laboratory of Behavioral Medicine, Southwest Foundation for Biomedical Research, San Antonio, Texas 78284 (A.M.C.)

INTRODUCTION

Although there has been substantial attention paid to behavioral development among many species of Old World Monkeys, there are few longitudinal studies that make direct and systematic comparisons between species from a similar data base. Consequently, differences in the goals of the study, observation techniques, and behavioral definitions render most literature reviews impractical except for a focus upon certain readily identifiable conditions (habitat, range, group composition, infant contact with mother, menarche, pregnancy, etc.). The three species in the present study are the focus of several studies of behavioral development. Struhsaker [1971] and Krige and Lucas [1975] provide a clear description of the changes in color and general pattern of neonate development for vervets. Reproductive maturation is described in Bramblett et al [1975] for both of the guenons. Bramblett [1980] documents age-correlated changes in activities of vervets from a cross-sectional study. Longitudinal records of individual rank indicate that dominance status is matriline-dependent among adult vervets [Bramblett et al, 1982]. Fairbanks and McGuire [1984] report shorter birth intervals and more surviving infants among high-ranking vervet females, an observation supported by observations of free-ranging animals [Whitten, 1983].

Longitudinal studies of Sykes' monkeys are few in number. Chalmers [1972] compared the first 3 months of life in vervet and Sykes' monkeys and observed that Sykes' infants were more reluctant to move away from their mother, a difference that Chalmers attributed to the Sykes' more arboreal habits.

In contrast to the Sykes' monkeys, there is a substantial and diverse literature on the development of the savannah baboon. Altmann [1980], Nash [1978], Ransom and Rowell [1972], and Rhine et al [1984] have described early social development in baboons. Cheney [1977] discusses the development of reciprocal alliances among immature baboons. Baboon physical development is well documented. The general stages of physical maturation are presented in Altmann et al]1981] and quantitative descriptions of baboon growth have been published [Glassman et al, 1984; Coelho, 1985].

In this paper, we make direct comparison between two guenon species, vervet and Sykes' monkeys, and the savannah baboon, in terms of rates of play, grooming, and aggression from birth to 87 months of age. The study was done in the controlled conditions of the laboratory, and an attempt is made to represent the information from the three species in as similar a form as possible.

METHODS

Subjects

The vervet and Sykes' groups were formed in 1966 by T.E. Rowell at Makerere University College, Kampala, Uganda [see Rowell, 1970, 1971]. They were imported to the University of Texas in November, 1968 (vervet group) and February, 1969 (Sykes' group). They were housed in chain link indoor and outdoor caging. The guenons were housed in outdoor cages (26 × 15 × 4 m for vervets and 7 × 15 × 4 m for Sykes'). Guenons were observed in their natal groups, comprised of all age and sex classes, kin and nonkin social units [Bramblett, 1978a]. Baboons were housed and observed in peer groups. Three fourths of the baboon infants were left with their mother for the first 90 days and then introduced into peer groups of 24 animals (12 males and 12 females). The remaining one fourth of the baboon sample were removed from their mother at birth and were human reared in a nursery until about 90 days of age when they were introduced to one of the peer groups described above. Four such peer groups are included in this study. The housing conditions of the baboons are described in detail elsewhere [see Coelho and Bramblett, 1981a], but a simplified description would be that they were housed in four peer-only groups in outdoor cages (15 × 7 × 8 m).

Data Tabulation and Analysis

Activities are summarized from 1,041.1 hr of focal animal tests on vervets (21 males and 21 females), 204.3 hr on Sykes' monkeys (seven males and seven females), and 2,816.4 hr on baboons (48 males and 48 females). Guenon figures are based upon a mixed longitudinal design over a period of 3 years. Baboon data are truly longitudinal, with each animal being sampled regularly from birth.

BEHAVIORAL MEASURES

Play among guenons was defined as a single category. It was scored as involvement in play activity in format [see Bramblett, 1978b, for a detailed explanation]. Definitions and characteristics of baboon response elements are discussed elsewhere [Hall and DeVore, 1965; Coelho and Bramblett, 1981b; Coelho and Bramblett, in press]. Baboon play in this study included the following behaviors: play face, enlist to play, grapple, and soft bite. Guenon aggression included bite, chase, cuff, supplant, fight, and take object from. Baboon aggression is a summation of charge, chase, supplant, hard bite, and pull. In all species, grooming refers to the specific act, groom.

The data for each animal were tabulated in 3-month intervals and expressed as a rate (bouts per hour). Rates for each individual were smoothed using running medians [Tukey, 1977]. Individual animals were then combined to compute mean rates for each species in 3-month intervals. Although the data extend beyond that age, a diminishing sample size results in truncation of the graphs at 87 months. For convenient cross species comparison, figures are organized by gender and activity.

RESULTS

Play has a virtually identical pattern in all three species (Fig. 1). In females, there is remarkably little difference, especially when one considers that these represent three different sets of animals with different surroundings (both physical and social). Play is a juvenile activity that declines with approaching puberty, especially among females. After puberty, males continue to play much more frequently than do females (Fig. 2). There is a particularly close correspondence between guenon and baboon females. It is important to note that vervet females reach adult rate of play by the beginning of their third year, almost 18 months earlier than the other two species. Male guenons differ from male baboons, the latter being frequent players. The juvenile pattern of play declines more slowly among baboons, until it finally

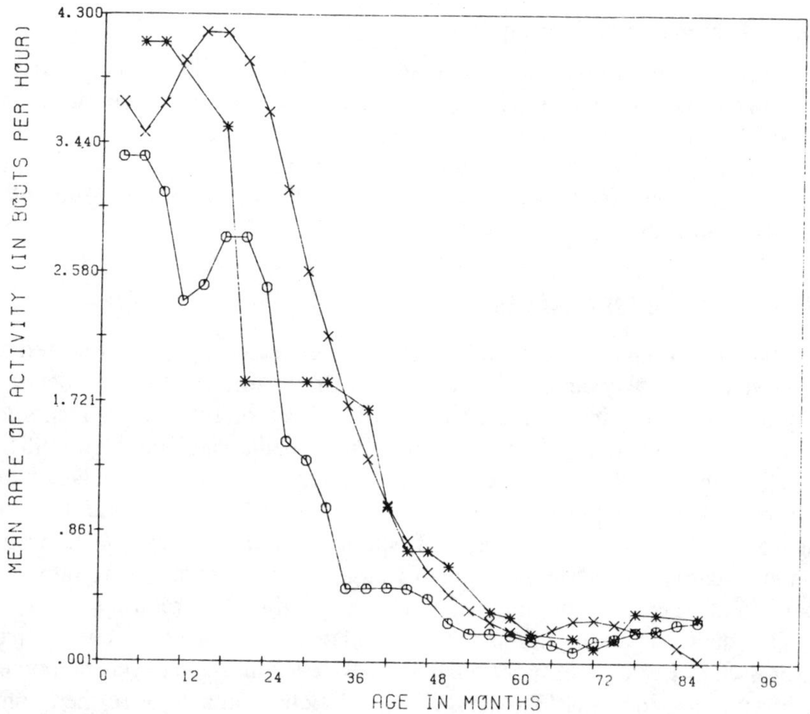

Fig. 1. Mean rates of playing by females as a function of the age of the performer for vervets (o), Sykes' (*), and baboons (X).

approaches the guenon adult level. Male guenons establish adult frequencies of play about the age of 4 years, but baboons do not show the adult rates until the age of 6.

Grooming represents a very different behavioral dimension. It is an adult, rather than a juvenile, activity. Guenon females exhibit a steady increase in grooming rates from infancy through 87 months (Fig. 3). At 7 years of age, they have reached adult frequencies and are an integrated part of their matrilines. Baboons, on the other hand, have undergone tubal ligations to prevent pregnancy, so the comparison of the later ages may be somewhat misleading. Generally, baboon females groom more frequently than guenons as infants, but the rates converge before the end of the second year. After that, all three species show comparable female rates. Male grooming includes some interesting contrasts (Fig. 4). Male baboons are relatively frequent groomers compared to guenons. Like play, grooming in baboons declines with age, but declines more rapidly. It has almost reached adult

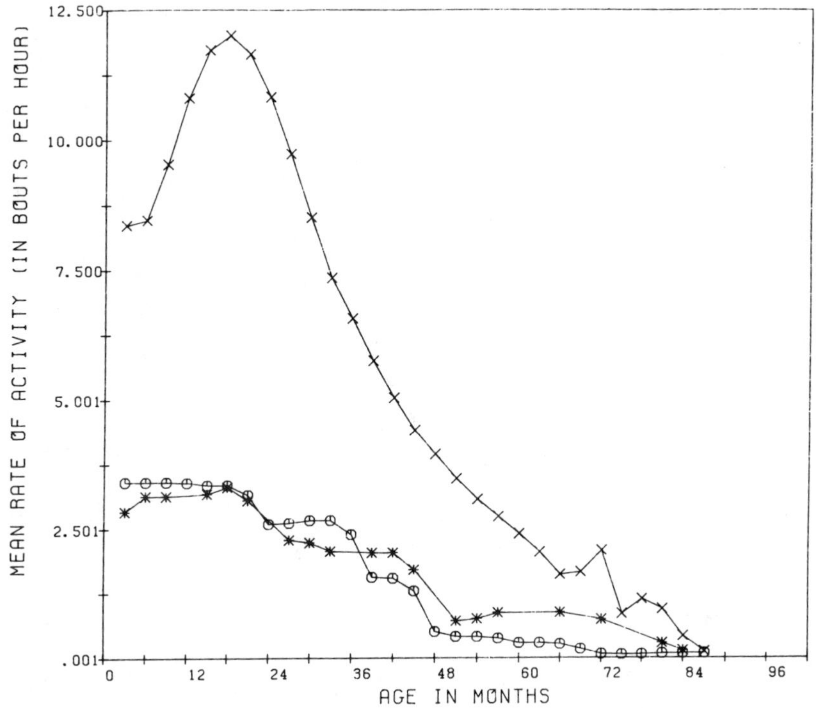

Fig. 2. Mean rates of playing by males as a function of the age of the performer for vervets (o), Sykes' (*), and baboons (X).

levels by the age of 3 years. There is a burst of activity near the age of 6 years, which subsides quickly to adult rates. Vervet males steadily increase their rates of grooming during their first 3 years, at which time they reach the adult pattern. Like baboons, vervet males exhibit a transient increase in grooming rates at the end of the fifth year. Sykes' monkey males groom less than the other two species until the sixth year, when there is a dramatic increase. Baboon and vervet males exhibit a marked decline in grooming activity after the sixth year.

Aggression illustrates a third behavioral dimension, more frequent in males than in females. Guenon females are basically similar (Fig. 5). Aggression rates are highest as they approach 2 years of age, but adult levels are only slightly lower. In contrast, baboon females are much more active than guenons but their peak in activity is very close to the same age as guenons. Their decline to adult levels is more extreme and dramatic, with adult rates reached at about 5 years of age. Vervet males show a pattern

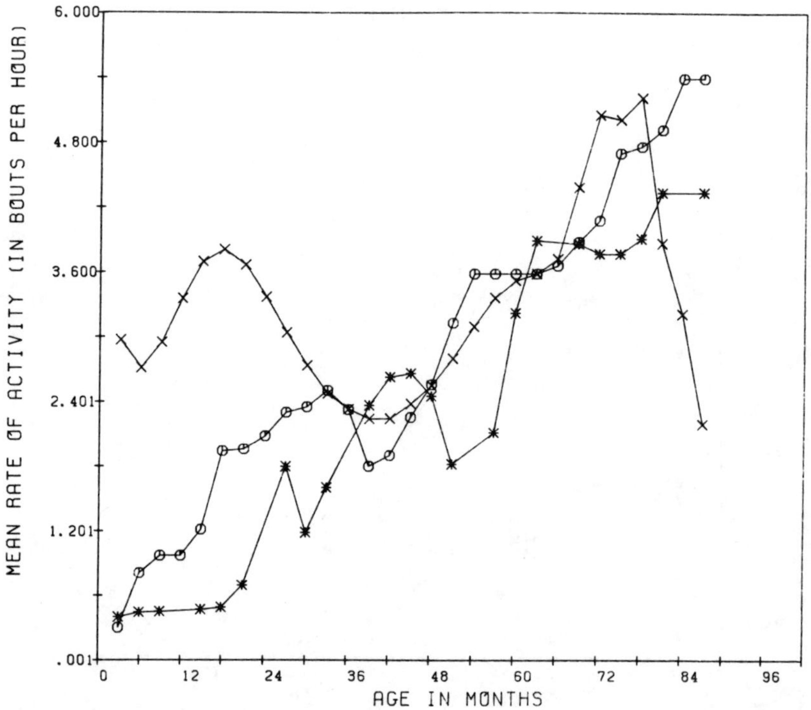

Fig. 3. Mean rates of grooming by females as a function of the age of the performer for vervets (o), Sykes' (*), and baboons (X).

reminiscent of females except that the maximum rates are reached near the age of 3 years, about a year later than in females (Fig. 6). Sykes' males are somewhat different. The decline in Sykes' male aggression does not occur until after the sixth year. Baboon males, as usual, are much more active as juveniles, but they reach adult rates that are similar to guenons by 4 years of age.

DISCUSSION

The three species present an interesting contrast in development, dimorphism, and social strucutre. Gartlan and Brain [1968] review the variation in social structure among vervets and assume that they show an adaptation intermediate between the more arboreal Sykes' monkey and an extreme terrestrial primate such as the patas monkey. Vervets are very abundant and widely distributed in Subsaharan Africa [Wolfheim, 1983]. They are most

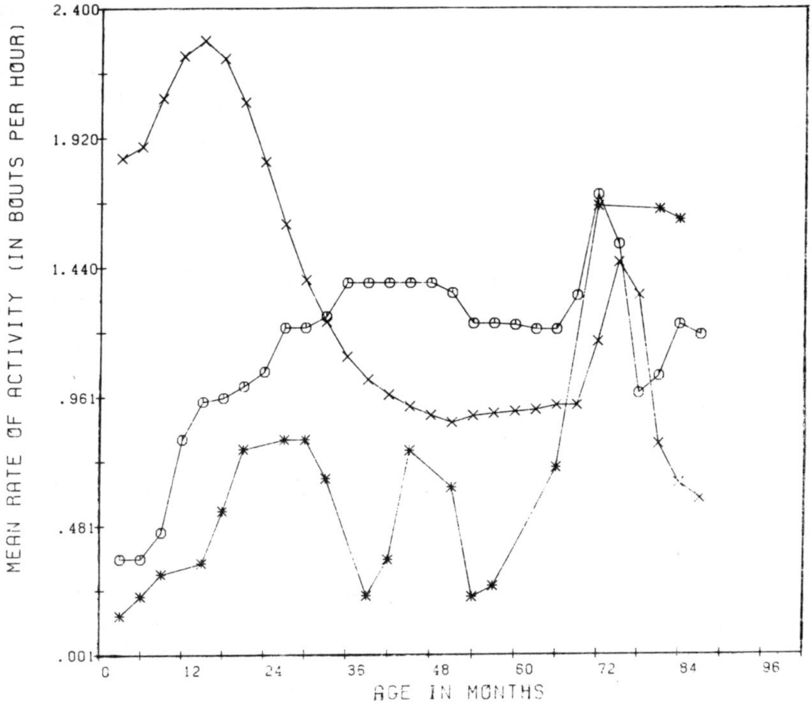

Fig. 4. Mean rates of grooming by males as a function of the age of the performer for vervets (o), Sykes' (*), and baboons (X).

often seen in woodland or forest edges near savannah or riparian woodland. They require both potable water and trees (or thickets) as sleeping sites. Sykes' monkeys have a discontinuous distribution that generally coincides with evergreen forests in Central, East, and Southeast Africa. Their primary habitat is the evergreen forests of Central Africa south of the Congo River. The Sykes' appears less tolerant of the higher temperature (especially the hot sun) that is characteristic of the vervet's woodland savannahs. Both guenons spend part of their foraging time on the ground. Both species are caught in traps that are placed on the ground.

Baboons, in contrast, are much larger bodied and primarily ground-living monkeys. They are inept climbers when compared to vervets [Bramblett, 1967] and can tolerate very poor quality, even brackish or alkaline, water supplies. Developmental differences in activities of these three species most likely represent varying adaptations to different niches along the savannah/woodland interface. The wide distribution of all three species reflects the

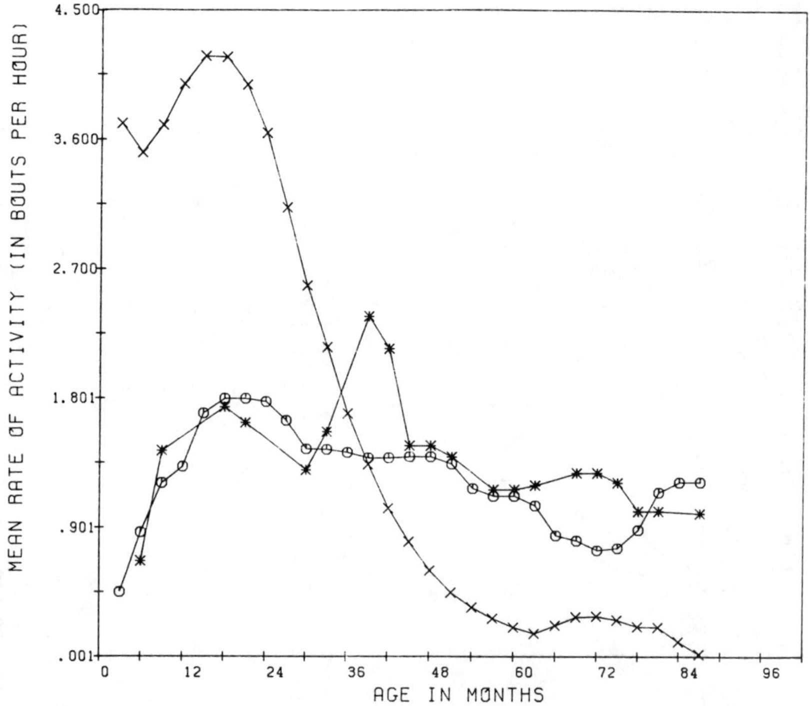

Fig. 5. Mean rates of aggression by females as a function of the age of the performer for vervets (o), Sykes' (*), and baboons (X).

extent of this habitat in the African continent. The relatively discontinuous distribution of the Sykes' monkey may reflect the less continuous nature of their primary habitat (the riverine woodland). Harvey and Clutton-Brock [1985] were unable to find general correlations between life history and ecology in a broad (but superficial) survey of life history variation in primates. However, we can observe functional relationships between behavior, life history, and ecology in each of our study species.

If one pushes generalizations to the extreme that we risk losing accuracy, we can describe some of the differences between the three species. The Sykes's are a sexually dimorphic species, with the male being slightly larger (14 to 20 lb) than the female. These males have larger canine teeth (and the associated complex of upper body muscles). Males develop a laryngeal sac at puberty that allows a male-specific call (a boom). Growth is slow, reaching apical closure of the canines and behavioral maturity near the age of 7 years. Social groups are small (eight to ten animals) and usually contain a

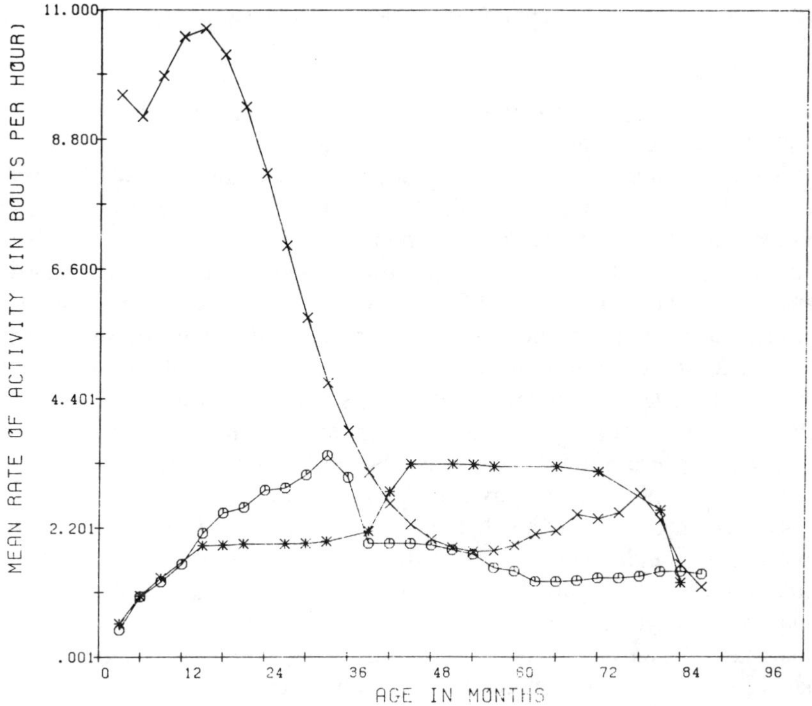

Fig. 6. Mean rates of aggression by males as a function of the age of the performer for vervets (o), Sykes' (*) and baboons (X).

single adult male, several adult females, and their associated followers. The major social organization of the group is that of related female matrilines. The adult male performs stereotypic roles of defense and display, but is not necessarily the most dominant animal in the group. The primary habitat for this species is arboreal, perhaps reducing the danger from ambush by a predator. Attacks by aerial predators may be relatively easy for a single adult male to deter by display, distraction, and if necessary, counterattack. Groups form by the fission of individuals, including the possible fission of male-female pairs. The pelage is dark and heavy (appropriate to cool nights and moderate days). A Sykes' monkey of any age cannot survive prolonged exposure to hot direct sunshine. The white collar breaks the animal's silhouette against the canopy, and like most forest mammals, the ears have white fringes that outline ear movements. Likewise, facial hairs with contrasting colors signal facial expressions. Sykes' monkeys adapt well to zoological gardens, but they reproduce slowly when compared to vervets and baboons [Else et al, 1985].

Vervets are smaller, less dimorphic monkeys (8 to 10 lb for adult females, and 10 to 14 lb for adult males). The pelage is lighter in color and shorter except for the cape. Vervet pelage can become quite heavy when exposed to periods of cold weather. Social groups are somewhat larger, often 20 animals or more. The group consists of multiple adult males and several adult females and their dependents. Social organization, especially dominance [Bramblett et al, 1982] and spacial relationships [Ehardt-Seward and Bramblett 1980; Lancaster 1971] strongly reflect female matrilines. Males are transients that attach themselves politically to several troops during the course of their lifetime. Males perform defensive male roles, including alarm and attack toward an intruder, but vigilance and highly structured alarms are the primary defense. Dispersal of males through the group and on the periphery of the group may reduce ambushes by predators that hide and wait for prey—a pattern more typical of terrestrial predators.

Savannah baboons are much larger in size and show increased sexual dimorphism. Adult males in East Africa range from 40 to 60 lb. Adult females are less than half that size. While the male canine tooth and fight complex is exaggerated in size, the female canine tooth is reduced, being less lethal than even the canine in the much smaller vervet female. The body build changes markedly. The baboon has relatively shorter limbs and shorter tail. The hand and foot show terrestrial adaptations. Body size has increased enough to make vervets, even adult vervet monkeys, targets for baboon predation. The hair is coarse and sparse when compared to guenon pelage. Even though the savannah baboons can live in hot, dry, grassland areas, they have a limited capability to rid themselves of metabolic heat from strenuous activity in the hot sun. Much of their adaptation to the savannah is behavioral. They avoid excess heat-generating activity when the temperature is too high and seek shelter from the sun when it is needed. Baboon social groups are large, often 40 or more individuals. A troop may contain many adult males and many adult females. Males are dispersed throughout the group, and a primary feature of baboon society is that females and immature animals fall under the protective shadow of a nearby fully adult male. Size dimorphism and male roles of protection and policing the group have produced a male-dominated society that is very different from that of the guenons.

Baboons are extremely dimorphic in body size and both genders are very active during the juvenile period. This is especially noticeable in the elevated play and aggressive activity of baboon males. With their rapid development of size dimorphism and high rates of juvenile activity, baboons are unique among these three species. Part of this elevation in juvenile interactions may reflect the peer environment, but baboons are clearly characterized by conspicuous and frequent juvenile play and aggression. Lancaster [1971] has pointed out that it may be relatively unimportant to a vervet female whether she is a more skilled combatant than other females in her group, since her dominance rank does not seem related to these qualities. The hypothesis is that play-fighting would be less important to those species (and genders) that acquire rank by social contact rather than display and aggression. Sykes' are more like baboons in maturation, requiring 7 years for full eruption of canine and apical closure among males. Females become pregnant with their first infant about the age of 4 years. Vervet males exhibit canine maturation about 5 years of age and females are pregnant just after 3 years of age. The current data are not adequate to distinguish between adult male activities, since the data end near the age of social maturity for both baboons and Sykes' monkeys.

The increased variance in most of the Sykes' graphs is a reflection of smaller sample size. Likewise, the relatively smooth baboon graphs reflect the large sample at each interval. The decline in grooming activity in baboons at 7 years of age is marked in both genders, and is unexpected. All baboon activity seems to be declining at that age, but without a longer longitudinal sample, the activity of fully adult males is not represented.

CONCLUSIONS

A mixed longitudinal design is used to compare age changes in play, aggression, and grooming among socially living vervets (*Cercopithecus aethiops*), Sykes' monkeys (*C. mitis alboqularis*), and savannah baboons (*Papio cynocephalus*). All subjects are immature and laboratory born. Although differences are evident in the rates of play in all species, play appears to develop more slowly and to persist in baboons. Female Sykes' monkeys play much less than do female vervets. Males are more aggressive than females in all three species, but Sykes' monkeys are especially dimorphic. Adult rates for aggression are evident for each species by the fourth year. Males of each species groom less frequently than do females. Although males achieve adult grooming rates by the end of the first year of life, female guenons attain adult rates between their fourth and fifth year. It is difficult to relate female baboons to this pattern, since their activity peaks at about 24 months of age, but they are frequent groomers all their lives.

ACKNOWLEDGMENTS

Support for the guenon colony came initially from the Wenner-Gren Foundation for Anthropological Research (Grant No. 2635). Additional support came primarily from the anthropological laboratories, the University Research Institute, and the College of Liberal Arts at the University of Texas. Baboon colony and observation were supported by funds from the National Institutes of Health (National Heart, Lung, and Blood Institute) Grants HL 15914 and HL 19362. Numerous University of Texas students have contributed to the guenon caretaking and data base (often without remuneration). Sharon S. Bramblett served as alternate caretaker and helped with the tabulation of data and manuscript preparation.

REFERENCES

Altmann J. (1980): "Baboon Mothers and Infants." Harvard University Press: Cambridge.

Altmann J, Altmann S, Hausfater G (1981): Physical maturation and age estimates of yellow baboons, *Papio cynocephalus*, in Amboseli National Park, Kenya. Am J Primatol 1:389–399.

Bramblett CA (1967): Pathology in the Darajani baboon. Am J Phys Anthropol 26:331–340.

Bramblett CA (1978a): Sex differences in the acqusition of play among juvenile vervet monkeys In Smith EO (ed): "Social Play in Primates." New York: Academic Press, 33–48.

Bramblett CA (1978b): Is the concept of control group valid? A quantitative comparison of behavior of caged baboon groups. Am J Phys Anthropol 49:217–226.

Bramblett CA (1980): A model for development of social behavior in vervet monkeys. Dev Psychobiol 13:205–223.

Bramblett CA, Pejaver LD, Drickman DJ (1975): Reproduction in vervet and Sykes' monkeys. J Mammal 56:940–946.

Bramblett CA, Bramblett SS, Bishop DA, Coelho AMJr (1982): Longitudinal stability in adult status hierarchies among vervet monkeys (*Cercopithecus aethiops*). Am J Primatol 2:43–51.

Chalmers NR (1972): Comparative aspects of early infant development in some captive Cercopithecines In Poirier FE (ed): "Primate socialization." New York: Random House, pp 63–82.

Cheney DL (1977): The acquisition of rank and the development of reciprocal alliances among free-ranging immature baboons. Behav Ecol Sociobiol 2:303–318.

Coelho AMJr (1985): Baboon dimorphism: Growth in weight, length and adiposity from birth to 8 years of age. In Watts ES (ed): "Nonhuman Primate Models for Human Growth and Development. New York: Alan R. Liss, Inc., pp 125–159.

Coelho AM, Bramblett CA (1981a): Effects of rearing on aggression and subordination in Papio monkeys. Am J Primatol 1:401–412.

Coelho AM, Bramblett CA (1981b): Interobserver agreement on a molecular ethogram of the genus Papio. Anim Behav 29:443–448.

Coelho AM, Bramblett CA Behavior of the genus Papio: Ethogram, taxonomy, methods and comparative measures. In Seth PK (ed): "Perspectives in Primate Biology," Vol 2. New Delhi: Today and Tomorrow Printers and Publishers (in press).

Ehardt-Seward C, Bramblett CA (1980): The structure of social space among a captive group of vervet monkeys. Folia Primatol 34:214–238.

Else JG, Eley RM, Suleman MA, Lequin RM (1985): Reproductive biology of Sykes' and Blue monkeys (*Cercopithecus mitis*). Am J Primatol 9:189–196.

Fairbanks, LA, McGuire MT (1984): Determinants of fecundity and reproductive success in captive vervet monkeys. Am J Primatol 7:27–38.

Gartlan JS, Brain CK (1968): Ecology and social variability in *Cercopithecus aethiops* and *C. mitis*. In Jay P (ed): "Primates: Studies in Adaptation and Variability." Holt, Rinehart and Winston, New York: pp 253–292.

Glassman DM, Coelho AM Jr, Carey KD, Bramblett CA (1984): Weight growth in savannah baboons: A longitudinal study from birth to adulthood. Growth 48:425–433.

Hall KRL, DeVore I (1965): Baboon social behavior: In DeVore I (ed): "PB." Holt, Rinehart and Winston, New York. 1965.

Harvey PH, Clutton-Brock TH (1985): Life history variation in primates. Evolution 39:559–581.

Krige PD, Lucas JW (1975): Behavioural development of the infant vervet monkey in a free-ranging troop. J Behav Sci 2:151–160.

Lancaster JB (1971): Play-mothering: The relations between juvenile females and young infants among free-ranging vervet monkeys (*Cercopithecus aethiops*). Folia Primatol 15:161–182.

Ransom TW, Rowell TE (1972): Early social behavior of feral baboons. In Poirier FE (ed): Nash LT (1978): The development of the mother-infant relationship in wild baboons (*Papio anubis*). Anim Behav 26:746–759. "Primate Socialization." New York: Random House, pp 105–144.

Rhine RJ, Norton GW, Westlund BJ (1984): The waning of dependence in infant free-ranging yellow baboons (*Papio cynocephalus*) of Mikumi National Park. Am J Primatol 7:213–228.

Rowell TE (1970): Reproductive cycles of two *Cercopithecus* monkeys. J Reprod Fertil 22:321–338.

Rowell TE (1971): Organization of caged groups of *Cercopithecus* monkeys. Anim Behav 19:625–645.

Struhsaker TT (1971): Social behaviour of mother and infant vervet (*Cercopithecus aethiops*). Anim Behav 19:233–250.

Tukey JW (1977): "Exploratory Data Analysis." Addison-Wesley: Reading.

Whitten PL (1983): Diet and dominance among female vervet monkeys (*Cercopithecus aethiops*) Am J Primatol 5:139–159.

Wolfheim JH (1983): Primates of the World. Seatle: University of Washington Press.

Comparative Behavior of African Monkeys, pages 81–97

5

Use of Space by Young African Green (*Cercopithecus aethiops* sp) and Patas (*Erythrocebus patas*) Monkeys in Captivity

Margaret R. Clarke, Jay R. Kaplan, Elizabeth C. Nicks, and Anna R. Grady

Department of Comparative Medicine, Wake Forest Medical Research Center, Winston-Salem, North Carolina 27103

INTRODUCTION

Patas monkeys (*Erythrocebus patas*) and African green monkeys (*Cercopithecus aethiops*) live sympatrically in many areas of their natural habitat in Africa. Although the two species are often found in proximity, patas live on the edges of grasslands and forest [Hall, 1965a], while the green monkeys inhabit these areas and extend into the forest and riverine areas [Booth, 1962; Gartlan and Brain, 1968; Hall, 1965b; Harrison, 1983; Lee, 1984; Struhsaker, 1967]. Patas and green monkeys are considered to be closely related species, and a green-patas hybrid has occurred in a laboratory situation [Matsubayashi et al, 1978]. There are, however, no reports of such a hybrid in the wild, although other species of *Cercopithecus* are known to produce feral fertile crosses [Butynski, 1982].

Green monkeys and patas are considered the most terrestrial of the *Cercopithecus-Miopithecus-Erythrocebus* genus group, with patas social groups spatially more dispersed and more terrestrial than green monkey groups [Hall, 1965a,b; Hall and Mayer, 1967; Gartlan, 1974]. Interactions between the two species have been characterized in the literature as "pacific" [Struhsaker and Gartlan, 1970], or at least as inconclusive [Hall, 1965b]. Patas and green monkeys do interact with other species in an obvious manner. Green monkeys are often aggressive to other green monkey groups

[Cheney, 1981], and have been observed being supplanted by *C. mitis* and *C. ascanius* [Booth, 1962; Rowell, 1969]. Patas have been observed being supplanted by baboons in their natural habitat [Struhsaker and Gartlan, 1970; Hall, 1965b] and by rhesus monkeys on the island colony of La Parguera in the Caribbean [Zucker, this volume]. Recent data from the field [Chism and Olson, personal communication] indicate that groups of patas and green monkeys could alternately supplant each other at a study site in Ghana, but that patas always supplanted green monkeys at a study site in Kenya. Green monkeys and patas thus exhibit no clear pattern of supplantation or avoidance, and species isolating mechanisms separating them are likely to be subtle. It is suggested here that it may be a differential use of habitat by these two species, manifested through a differential use of space.

If animals really do use space differently, it should be testable by housing them in a condition that is identical yet foreign to both species [Rowell, 1969]. Thus, patas and green monkeys housed in captivity should exhibit differential use of space if this is a real difference in their natural habitat. We should be able to determine how they use vertical and horizontal space, and if it differs according to ambient conditions. We should also be able to determine if individuals preferentially orient themselves toward members of their own species, and if these spatial arrangements are reinforced by social behavior. If they do use space differently in captivity, we may be able to understand how this relates to their behavior in their natural habitats, and how it may serve to separate these species in the wild.

METHODS

A 12-month study on use of space was conducted on young patas and green monkeys housed in an outdoor corral at the Research Center of the Bowman Gray School of Medicine of Wake Forest University, Winston-Salem, North Carolina. The number of patas monkeys (*E. patas*) in the study group varied between nine and 15 individuals, and they were housed with seven African green monkeys (*C. aethiops aethiops, C. a. sabeus*, and *C. a. aethiops* × *C. a. sabeus*). The monkeys' ages were between 6 months and 2 years at the beginning of the data collection period, at which time some of the subjects had been living together for 4 months. The corral consisted of a ⅓-acre enclosure (35 × 30 m) surfaced with coarse gravel and equipped with two 2-story wooden platforms (Fig. 1). There was also an inside area with perches that was heated during the winter months. Food and water were available both inside and outside.

Data were collected and analyzed as follows.

Fig. 1. Outdoor enclosure.

Patas/Green Monkey Spacing

In order to determine how the animals used space and to determine if the species were spatially separated, instantaneous scan samples were done between 7:30 AM and 8:30 PM from August, 1981 through July, 1982. Each observation period consisted of three scan samples at 5-min intervals. The species and location of each monkey in the corral were noted. Individuals were counted as being on platform 1, on platform 2, under platform 1, under platform 2, inside, or on the ground away from the platforms. The three samples of each observation period were averaged to represent a mean percentage of animals of either species at any one location, which was then treated as a single observation. These were combined on a monthly basis by time of day and by temperature (see Table I for definitions). Additionally, each location (platform 1, platform 2, or inside) was scored as "primarily patas," "primarily green monkeys," or "mixed." Thus, using a two-thirds criterion, 66.7% of the animals at any one location had to be of one species for that observation to be counted as "primarily" of that species. These observations were also averaged for each month. The data for January and February were combined due to few observations.

Patas/Green Monkey Interaction

In order to determine if spatial differences at the species level were related to aggression or affiliation between species members, status matrices [after Bramblett, 1981] for aggressive interactions, grooming bouts, and frequencies of being in contact with other animals were analyzed for 3-month periods

TABLE I. Categories of Analysis

Designation	Definition
Time	
AM	7:30 AM to 10:30 AM
Mid	10:31 AM to 2:00 PM
PM	2:01 PM to 8:30 PM
Temperature (°F)	
Low	Less than 45
Mild	45–70
High	More than 70
Location	
Inside	Indoor area
Platform 1	On or near platform 1
Platform 2	On or near platform 2
Down	On the ground away from the platforms
Up	On the platforms and/or on inside perches

from August, 1981 through July, 1982. These data were part of a long-term study on interaction patterns and physical activity of patas and green monkeys. The data in these matrices were derived from two 15-min focal samples per animal per week for each 3-month period.

Green Monkey Spacing

In order to determine if greens housed *without* patas used space differently than those housed *with* patas, scan samples identical to the patas/green monkey spacing samples were done on two adjacent ⅙-acre corrals of adult and juvenile green monkeys. These were scored as "on the ground," "indoors," or "on or near the platform." They were taken two or three times a day during July, 1982, and the month's data were combined.

RESULTS

Large proportions of green monkeys were more frequently found off the ground and were more frequently found indoors than were patas monkeys. Monthly averages are shown for patas and greens "inside" in Figure 2 (green $\bar{X}$ = 62%, mean range = 40–80%; patas $\bar{X}$ = 35%, mean range = 10–60%), and monthly averages for patas and greens "off the ground" are shown in Figure 3 (green $\bar{X}$ = 84%, mean range = 75–90%; patas $\bar{X}$ = 71%, mean range = 65–80%). With the exception of the combined average for January/February "inside," the averages for the green monkeys were higher than for the patas (sign test, $p = .006$).

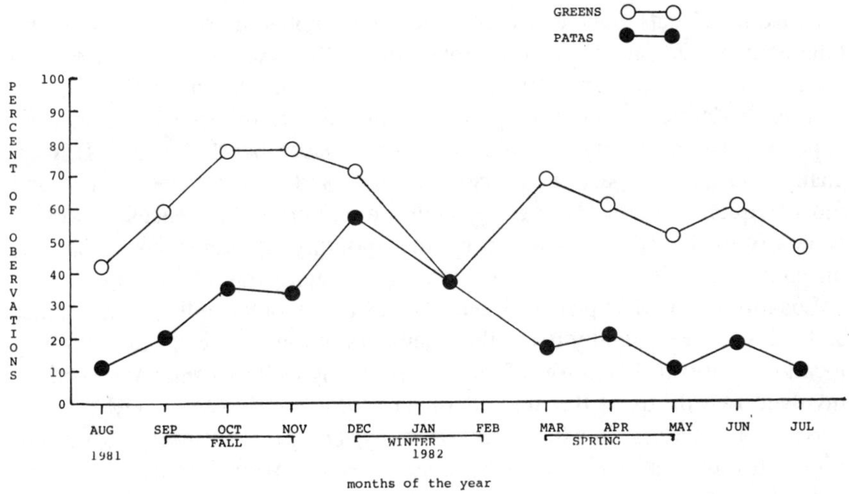

Fig. 2. Monthly average of percentage of patas and greens located inside.

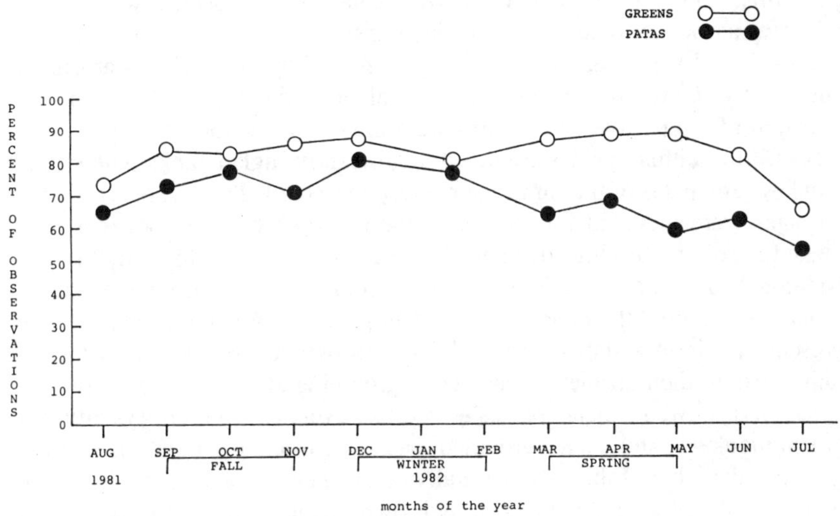

Fig. 3. Monthly average of percentage of patas and greens located off the ground.

Seasonal effects on the use of space were apparent when analyzed by temperature. During the lowest temperatures, there was no difference as to which species was found "inside" (Fig. 4) or "off the ground" (Fig. 5). During moderate and high temperatures, the species differences again were apparent, except during the coldest months (December–February). Diurnal changes in use of space were also analyzed, and both species were found more frequently on the ground later in the day. Diurnal differences, however, generally reflected the pattern of the monthly averages, with species differences in use of space diminishing during temperature extremes.

Locations scored as primarily one species or the other indicated that patas and greens were spatially near other members of their same species. For any location (platform 1, platform 2, inside), as many as 94% of the observations involved two thirds of the animals of one species only ($\bar{X}$ = 63%). There was also a difference in where green monkeys and patas groups were located. Figure 6 indicates that groups "primarily patas" were infrequently found indoors, while, except during winter months, groups "primarily green monkey" indoors comprised 45% or more of the observations. Mixed groups were most common indoors during the winter months when it was colder outside. Figure 7 indicates the opposite patterns for species groupings on the platforms. "Primarily green monkey" groups were infrequently found on the platforms except during the winter months when the ground was cold or snowy. Otherwise, "primarily patas" or mixed groups were found on platforms. Thus, patas associated with patas on the decks, while green monkeys associated with green monkeys inside.

This difference in use of space was reinforced by affiliative interactions, but not by aggressive interactions. Dominance hierarchies based on the direction of aggressive interactions were almost linear for all four 3-month blocks, but neither species occupied significantly higher ranks when green monkeys and patas of the same age range were included (see Table II). When all patas were included for the last two quarters, the green monkeys were ranked significantly higher than patas. It is important to note that the additional four patas were 6 to 12 months younger than those in the original group, and entered the hierarchy below all patas and green monkeys already present. Their low status was probably attributable to the age difference rather than to their species. Measures of grooming and in-contact (affiliative behaviors) showed a clear species effect, however. Green monkeys groomed green monkeys and patas groomed patas significantly more than either groomed the other (Table III), and individuals of each species were in contact more frequently with a member of their own species (Table IV). The overall rate of grooming within species was 0.5 per hr, and between species it was

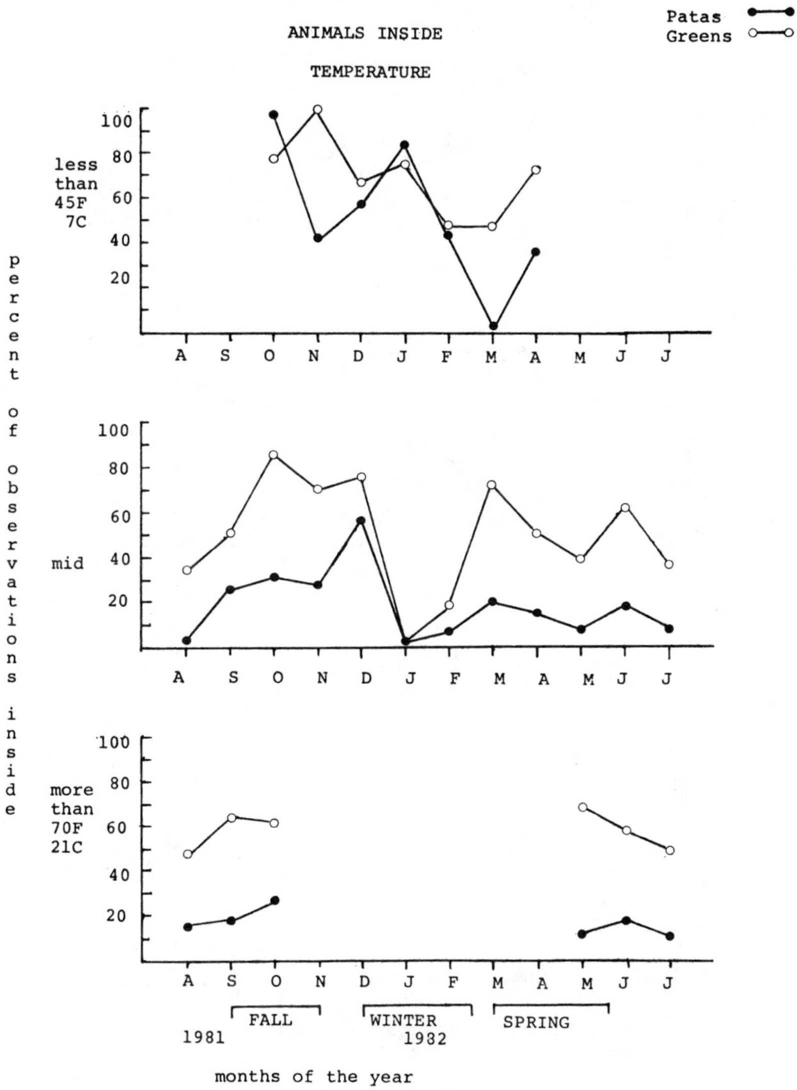

Fig. 4. Monthly average by temperature of percentage of patas and greens located inside.

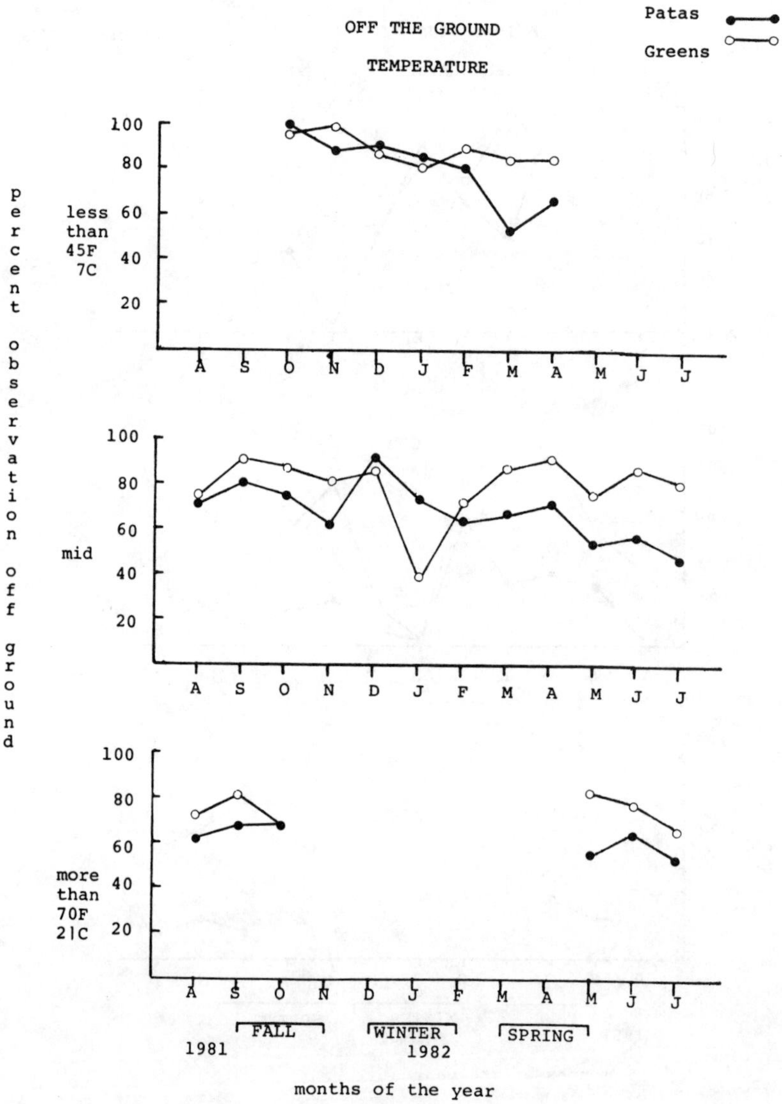

Fig. 5. Monthly average by temperature of percentage of patas and greens found off the ground.

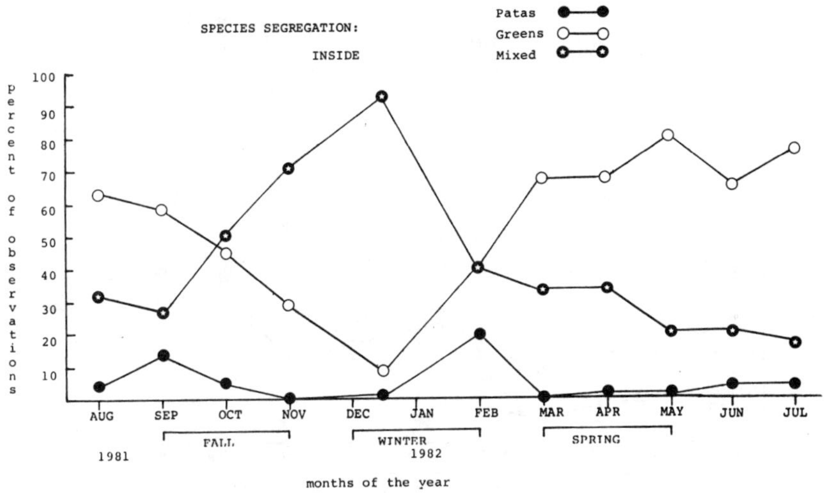

Fig. 6. Average monthly percentage of observations that animal groupings inside the sleeping area were over two-thirds patas, two-thirds green, or were mixed.

0.1 per hr. Rate of contact showed a similar distribution, being 3.5 per hr within species and 0.6 per hr between species.

The spacing data on the green monkeys living without patas were all taken when the ambient temperature was above 70°F, so these data were compared with observations of spacing by green monkeys living with patas when the ambient temperature was also above 70° (see Figs. 3 and 4). Green monkeys living without patas did not differ in the frequency of being found "off the ground," but did differ in the frequency of being found indoors. In the mixed species corral, 70 to 85% of the green monkeys were found "off the ground," while juvenile green monkeys without patas were found "off the ground" 80% of the time. Only 35% of juvenile green monkeys living without patas were found "inside" during observations, while 50 to 57% of the juvenile green monkeys living with patas were found "inside."

DISCUSSION

The results showed that young patas and African green monkeys did use space differently when housed together in a corral situation. The green

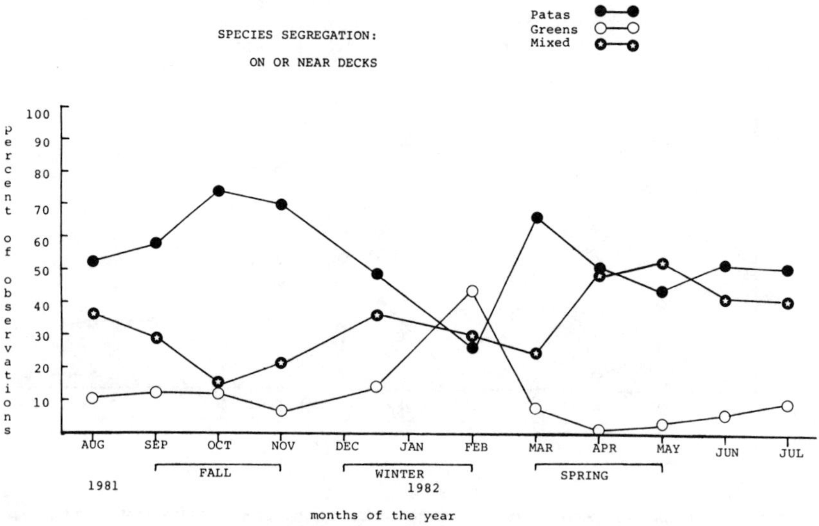

Fig. 7. Average monthly percent of observations that animal groupings on or near the decks were over two-thirds patas, two-thirds greens, or were mixed.

TABLE II. Group Size and Mann-Whitney U Scores for Dominance Ranks of Patas vs Greens

Quarter	Greens n_1	Patas n_2	Mann-Whitney U	Significance
1	7	9	49	n.s.
2	7	11	46	n.s.
3	7	11	49	n.s.
4	7	11	48	n.s.
With additional four young patas				
3	7	15	49	$p < .05$
4	7	15	49	$p < .05$

monkeys were more frequently found off the ground, and were more frequently found inside as individuals. In addition to this difference in vertical spacing, green monkeys and patas also appeared to align themselves spatially with members of the same species. Green monkeys were more frequently found in all (or almost all) green monkey groups, and patas were found in all (or almost all) patas groups. Patas groups tended to cluster on the platforms while green monkeys clustered indoors. From these data it was not

TABLE III. Grooming Scores and Rates

	Aug-Sept-Oct,'81	Nov-Dec,'81; Jan,'82	Feb-Mar-Apr,'82	May-Jun-Jul,'82	Year total
Quarterly scores[a]					
Green-green	27.6	18.3	18.4	24.6	89
Green-patas	3.0	4.0	1.8	4.0	13
Patas-green	1.7	1.0	1.9	2.5	7
Patas-patas	6.4	3.6	6.1	7.1	23
					$X^2 = 47.75$ sig. $p < .01$
					Mean
Quarterly rates/hr per individual					
Within species	0.6	0.3	0.4	0.5	0.5
Between species	0.1	0.1	0.1	0.1	0.1

[a]Data represent raw frequencies adjusted for number of potential recipients of grooming in each species during each quarter, and then adjusted to equalize number of observation periods.

TABLE IV. In-Contact Scores and Rates

	Aug-Sept-Oct,'81	Nov-Dec,'81; Jan,'82	Feb-Mar-Apr,'82	May-Jun-Jul,'82	Year sum
Quarterly scores[a]					
Green-green	120.3	134.0	83.9	113.9	452.1
Green-patas	13.1	10.4	5.6	11.5	40.6
Patas-green	16.2	19.6	22.2	33.0	91.0
Patas-patas	56.1	1.8	111.5	55.0	314.4
					$X^2 = 445.6$; $p < .001$
					Mean
Quarterly rates/hr per individual					
Within species	3.1	4.1	3.9	2.8	3.5
Between species	0.6	0.6	0.5	0.8	0.6

[a]Data represent raw scores adjusted for number of potential "contactees" in each species during each quarter, and then adjusted to equalize number of observation periods.

possible to determine if or how animals clustered, by species, when they were on the ground away from the platforms.

Species spatial orientation was reinforced by interspecific and intraspecific interactions. In the corral situation, patas and green monkeys did interact, but it was significant that affiliative interactions (grooming, in-contact) within species were much higher than between species, indicating that they both preferred members of their own species. Although green monkeys had a higher overall rate of contact and grooming than patas, patas still groomed patas significantly more than they groomed green monkeys. In the wild, patas and green monkeys have not been reported to interact in an affiliative manner, but apparently are limited to occasional aggression between species, or passive coexistence. The fact that the corral animals were all juveniles and were confined to a corral may account for the small amount of observed interspecific affiliative behavior.

Species distinctions in patterns of aggressive interactions were not as distinct. Although a relatively stable linear hierarchy based on aggression was evident, the seven green monkeys were distributed among the first 13 ranks. In terms of observed aggression, patas were not clearly dominant over green monkeys, or vice versa. It is possible that the interspecific interactions involved more subtle communication than overt aggression. Patas are larger than green monkeys, even as juveniles, and larger primates tend to supplant smaller ones in feral conditions [Struhsaker, 1981]. There were also more patas than green monkeys in the corral (more than twice as many by the end of the study), and Chism and Olson [personal communication] reported that larger groups of free-ranging patas were more frequently observed supplanting green monkeys than were the smaller groups of patas at their study sites. The comparison of use of space by green monkeys living without patas suggests that patas are the dominant species. Although the green monkeys living without patas were found on the platforms as frequently as green monkeys living with patas, the green monkeys without patas were much less frequently found indoors than the green monkeys living with patas. Thus, green monkeys in the mixed-species corral might have been indoors to avoid patas. This inferred dominance of patas over green monkeys would also mean that patas chose to be on the ground, and that they were not simply on the ground rather than on the platforms because they were being excluded or supplanted.

This differential use of space reinforced by species interactions reflects patterns seen in the natural habitat. Green monkeys, who were more frequently up on the platforms than patas, are also more arboreal than patas in the wild. Poirier's [1972] estimate that St. Kitt's vervets spent more than 75% of their day in trees is within the range of this study's findings, and Hall [1965a] and Struhsaker [1967] also reported that green monkeys were more arboreal. Use of trees or ground for traveling by both species is obviously limited to some extent by what is available, but both species do utilize grassland. In riverine woodland in Kenya, green monkeys have been observed traveling on the ground when they could have traveled in the trees [Chism and Olson, personal communication]. Patas, however, generally do not enter the riverine woodland, a habitat more commonly associated with green monkeys. Although both species may be more arboreal than previous reports have suggested, the continued impressions from the field and from this study are that green monkeys are more arboreal than patas.

Given the opportunity, then, patas and green monkeys do segregate themselves by species, both in captivity and in their natural habitat. This differential use of space and preference for intraspecific affiliation can disappear, however, when conditions are particularly unusual. One type of unusual condition is extreme temperature. During the coldest winter months, green monkeys and patas tolerated close proximity in the indoor enclosure, and, when outside, congregated on the platforms when the corral surface was cold (if not covered with snow). During the summer months they also tolerated each other on or near the platforms during the higher temperatures when they were presumably seeking shade. Thus, spatial separation of the two species diminished when they were avoiding extreme temperatures.

A second unusual condition is high spatial density. A different pattern of interspecific affiliative interactions was observed in a mixed group of green monkeys and patas who were matched for age-sex composition with the animals in the present study, but who were housed in a much smaller area (37.5 m^2). The rate of being in contact with another animal was higher for both between and within species contact in the "crowded" condition than in the corral. Intraspecific contact in the crowded condition was higher by 70% (3.5/hr corral vs 5.9/hr crowded), while interspecific contact in the crowded condition was increased by over 500% (0.6/hr corral vs 3.2/hr crowded). Rates of intraspecific grooming were 40% lower in the crowded condition (0.5/hr corral vs 0.3/hr crowded), while rates of interspecific grooming in the crowded condition were twice those in the corral (0.1/hr corral vs 0.2/hr crowded). Thus, rates of interspecific interactions were higher for animals housed in closer quarters. Data from these same two groups for an earlier time period indicated the same trends [Zucker, 1981]. Thus, for the most part, animals oriented themselves spatially to members of their own species, and interacted preferentially with conspecifics except under unusual circumstances, such as extreme temperature or crowding.

As patas and green monkeys do not appear to compete directly for resources in the wild, the differential use of space and intraspecific preferences indicated by this study may be keeping patas and green monkeys apart in their natural habitat. As they do interact in unusual circumstances, it is not surprising that patas and green monkeys do overlap on occasion in the wild without conflict. The loose spatial organization of both species should contribute to tolerance [Kaplan, this volume], but potential conflict between the two species is probably also reduced by green monkeys' increased utilization of available vertical space.

How does this spatial relationship of patas and green monkeys relate to other sympatric species of the *Cercopithecus-Erythrocebus-Miopithecus* genus group? Patas, despite their frequent utilization of trees when trees are available, are still considered the most terrestrial of the species in this group and can supplant green monkeys, at least when patas are more numerous. Green monkeys, while often found on the ground, may also be kept out of the forest by larger *Cercopithecus*. Green monkeys would compete with *C. mitis* if the green monkeys moved into the deeper forest, and *C. mitis* have been reported to supplant green monkeys [Booth, 1962; Moreno-Black and Maples, 1977]. Gartlan and Brain [1968] reported that *C. mitis* move higher in the canopy when they encounter green monkeys, which they interpreted as meaning that green monkeys were dominant. Since *C. mitis* are reported as being totally arboreal [Aldrich-Blake, 1970; Rudran, 1978], moving higher in the trees may not necessarily be a sign of being supplanted by green monkeys, but a partitioning of space in response to a semiarboreal species.

Cercopithecus ascanius also inhabit gallery forest contiguous with the grasslands inhabited by green monkeys [Rowell, 1969]. It may be possible that *C. ascanius* are preventing green monkeys from entering the forest. In addition, some subspecies of *C. ascanius* have been reported as partially terrestrial [Buxton, 1952], which would increase the chance that they could compete with green monkeys. Other species of *Cercopithecus* reported as utilizing the ground are *C. campbelli* (Lowe's guenon), who come to the ground occasionally to play or travel [Bourliere et al, 1970; Hunkeler et al, 1972], *C. mona*, who are occasionally seen on the ground [Struhsaker, 1969], and *C. l'hoesti*, who are often seen on the ground [Struhsaker, 1969; Rudran, 1978]. Green monkeys, while apparently still more terrestrial than these three species, are not reported to compete with them. It would appear that green monkeys are being kept out of the most open habitat by the larger patas and are also kept from entering the deep forest by *C. mitis,* and possibly also by *C. ascanius*. Neither patas nor green monkeys would come into contact with the totally arboreal species *C. diana* [Kawata, 1980], *Miopithecus talapoin* [Rowell, 1973], *C. nictitans,* and *C. cephus* [Struhsaker, 1969] except under unusual circumstances.

To what extent can the use of space by semiterrestrial patas and green monkeys be compared with the more arboreal *Cercopithecus* species? One comparison might be the different uses of space by semiterrestrial, mostly arboreal, and totally arboreal species. In this case, a comparison of semiterrestrial patas and green monkeys with mostly arboreal *C. ascanius, C. campbelli, C. l'hoesti,* and with the other totally arboreal *Cercopithecus* species would be appropriate, particularly because they are closely related phylogenetically. Another comparison, however, would be the use of space by different species that utilized the same strata of the habitat. In order to make this comparison, patas and green monkeys should be compared with other semiterrestrial species of Old World monkeys. Use of space could then be interpreted as a species difference due to different social structure or feeding patterns, rather than as a utilization of a different stratum of the forest. Patas could be compared with the more terrestrial species, such as drills, mandrills, and some *Macaca* (*M. sylvana, M, fuscata*), while green monkeys could be compared with the more arboreal rhesus macaques [Lindburg, 1971; Teas et al, 1980], toque macaques [Dittus, 1980], and forest living baboons [Rowell, 1966]. Comparisons of patas and baboon social structure [Hall, 1965a] and habitat utilization [Hamilton, 1981] already indicate that these species are very different despite their common terrestriality. Differences in use of space in the same habitat may illuminate species differences not readily apparent from other types of analyses.

Thus, patas and green monkeys, when confined in a corral situation, do exhibit different use of space, which appears to reflect their use of space in their natural habitat. Species separation corresponds to differences in spatial orientation, as well as in interaction patterns. Behavioral mechanisms for species separation do diminish under stress, such as crowding or extreme temperature, but it appears that these species do place themselves preferentially with members of their own species, and that this use of space is reinforced by preferential affilitative interactions with conspecifics. They appear, then, to continue to exhibit behavioral tendencies that may have been involved in the separation of these closely related species.

CONCLUSIONS

1. Juvenile patas and green monkeys housed in an outdoor corral utilized available space differently. Patas were found on the ground and outdoors more frequently than were green monkeys.
2. Members of each species preferentially associated with members of their own species, both in spatial proximities and social interactions.
3. This spatial separation diminishes under stress, such as crowding or extreme temperatures.
4. Green monkeys living without patas utilized all parts of their corral environment, which suggests that patas have priority of access to preferred locations.
5. These patterns reflect these species' use of space in their natural habitats.

ACKNOWLEDGMENTS

We thank H. Strickland and E. Zucker for additional observations, L. Rudel for permission to collect and utilize data from the green monkey groups in his studies, and S. Staley and F. Lusso for computer assistance. This research was supported by National Research Service Award #HL07115 to MRC and NIH Grant #R01H626562 to JRK.

REFERENCES

Aldrich-Blake FBG (1970): Problems of social structure in forest monkeys. In Crook JH (ed): "Social Behavior in Birds and Mammals." New York: Academic Press, pp 79–101.

Booth C (1962): Some observations on behavior of *Cercopithecus* monkeys. Ann NY Acad Sci 102:477–487.

Bourliere F, Hunkeler C, Bertrand M (1970): Ecology and behavior of Lowe's guenon (*Cercopithecus campbelli* lowei) in the Ivory Coast. In Napier J, Napier P (eds): "Old World Monkeys: Evolution, Systematics, and Behavior." New York: Academic Press, pp 297–350.

Bramblett CA (1981): Dominance tabulation: Giving form to concepts. Behav Brain Sci 4:435–436.

Butynski TM (1982): Harem male replacement and infanticide in the blue monkey (*Cercopithecus mitis* stuhlmanni) in the Kibale Forest, Uganda. Am J Primatol 3:1–22.

Buxton AP (1952): Observations on the diurnal behaviour of the redtail monkey (*Cercopithecus ascanius* schmidti matschie) in a small forest in Uganda. J Anim Ecol 21:25–58.

Cheney DL (1981): Intergroup encounters among free-ranging vervet monkeys. Fo Primatol 35:124–146.

Dittus WPJ (1980): The social regulation of primate populations: A synthesis. In Lindburg DG (ed): "The Macaques: Studies in Ecology, Behavior and Evolution." New York: Van Nostrand Reinhold, pp 263–286.

Gartlan JS (1974): Adaptive aspects of social structure in *Erythrocebus patas*. Symposium of the Fifth Congress of the International Primatological Society, pp 161–171.

Gartlan JS, Brain CK (1968): Ecology and social variability in *Cercopithecus aethiops* and *C. mitis*. In Jay PC (ed): "Primates: Studies in Adaptation and Variability." New York: Holt, Rinehart and Winston, pp 253–292.

Hall KRL (1965a): Behaviour and ecology of the wild patas monkey, *Erythrocebus patas*, in Uganda. J Zool 148:15–87.

Hall KRL (1965b): Ecology and behavior of baboons, patas, and vervet monkeys in Uganda. In Vagtborg H (ed): "The Baboon in Medical Research." Austin: University of Texas Press, pp 43–61.

Hall KRL, Mayer B (1967): Social interactions in a group of captive patas monkeys (*Erythrocebus patas*). Folia Primatol 5:213–236.

Hamilton WJ III (1982): Baboon sleeping site preferences and relationships to primate grouping patterns. Am J Primatol 3:41–53.

Harrison MJS (1983): Territorial behavior in the green monkey, *Cercopithecus sabaeus:* Seasonal defense of local food supplies. Behav Ecol Sociobiol 12:85–91.

Hunkeler C, Bourliere F, Bertrand M (1972): Le comportement social de la mone de lowe (*Cercopithecus campbelli* lowei). Folia Primatol 17:218–263.

Kawata K (1980): Notes on the comparative behavior in three primate species in captivity. Zool Garten N.F. 50:209–224.

Lee PC (1984): Ecological constraints on the social development of vervet monkeys. Behaviour 91:245–262.

Lindburg DG (1971): The rhesus monkey in north India: An ecological and behavioral study. In Rosenblum LA (ed): "Primate Behavior" Volume 2. New York: Academic Press, pp 1–106.

Matsubayashi K, Hirai M, Watanabe T, Ohkura Y, Nozawa K (1978): A case of patas-vervet hybrid in captivity. Primates 19:785–793.

Moreno-Black G, Maples WR (1977): Differential habitat utilization of four Cercopithecidae in a Kenyan forest. Folia Primatol 27:85–107.

Poirier FE (1972): The St. Kitts green monkey (*Cercopithecus aethiops sabaeus*): Ecology, population dynamics, and selected behavioral traits. Folia Primatol 17:20–55.

Rowell TE (1966): Forest-living baboons in Uganda. J Zool Soc Lond 149:344–362.

Rowell TE (1973): Variability in the social organization of primates. In Morris D (ed): "Primate Ethology." New York: Anchor Books, pp 219–235.

Rowell TE (1973): Social organization of the wild talapoin monkeys. Am J Phys Anthropol 38:593–598.

Rudran R (1978): Socioecology of the blue monkey (*Cercopithecus mitis* stuhlmanni) of the Kibale Forest, Uganda. Smithsonian Contrib Zool No. 249. Washington, D.C.: Smithsonian Institution Press.

Struhsaker TT (1967): Behavior of vervet monkeys and other Cercopithecines. Science 156:1197–1203.

Struhsaker TT (1969): Correlates of ecology and social organization among African Cercopithecines. Folia Primatol 11:80–118.

Struhsaker TT (1981): Polyspecific associations among tropical rainforest primates. Z Tierpsychol 57:268–304.

Struhsaker TT, Gartlan JS (1970): Observations on the bahaviour and ecology of the patas monkey (*Erythrocebus patas*) in the Waza Reserve, Cameroon. J Zool 161:49–63.

Teas, J, Richie T, Taylor H, Southwick CH (1980): Population patterns and behavioral ecology of rhesus monkeys (*Macaca mulatta*) in Nepal. In Lindburg DG (ed): "The Macaques: Studies in Ecology, Behavior, and Evolution." New York: Van Nostrand Reinhold, pp 247–262.

Zucker EL (1981): Social interactions of patas monkeys and African green monkeys in interspecific groups: Spatial influences. Paper presented at the 61st Meeting of the Western Psychological Association.

Comparative Behavior of African Monkeys, pages 99–125

6

Interspecies Interactions Between Free-Ranging Patas (*Erythrocebus patas*) and Rhesus (*Macaca mulatta*) Monkeys

Evan L. Zucker

Department of Psychology, Loyola University, New Orleans, Louisiana 70118

INTRODUCTION

Interactions between members of different species have drawn the attention and interest of researchers working in natural, seminatural, and captive environments. This interest has persisted over time, as reports of macaque-baboon interactions in captivity can be historically traced as far back as Hamilton's 1914 publication [see Erwin and Zucker, this volume].

The number of species whose members have been observed to interact with members of other species is quite large. Numerous published field studies have included sections dealing with "interspecies interactions," and often include mention of the subjects' responses to the observer(s). For example, Hall's [1965] pioneering work with patas monkeys in Uganda included brief mention of interactions with (or reactions to) baboons and vervets. A survey of some of the published reports that have specifically examined interspecific interactions in the wild indicates interactions among Malayan monkeys [Bernstein, 1967, 1968], Indian monkeys [Singh and Sen, 1977–78], Hong Kong monkeys [Southwick and Southwick, 1983], and monkeys in North Africa [Nagel, 1973], West Africa [Gartlan and Struhsaker, 1972], East Africa [Rose, 1977; Samuels and Altmann, 1986; Moreno-Black and Maples, 1977], and Central Africa [Aldrich-Blake, 1968; Hayashi, 1975; Struhsaker, 1981; Struhsaker and Leyland, 1979].

Descriptions of interspecific interactions of New World monkeys in the wild have been provided by Klein and Klein [1973], Izawa [1975], and Bernstein et al [1976]. Interspecific interactions in captivity have been reported for spider and capuchin monkeys [Maple and Westlund, 1975] and between squirrel monkeys and uakaries in a seminatural habitat [Abordo et al, 1975].

Much more interspecific "diversity" is possible under laboratory conditions, and is often the topic of investigation. Examples of such diversity include rhesus monkeys and baboons [Maple and Lawson, 1975; Maple et al, 1976], patas and crab-eating monkeys [Baker and Preston, 1973; Seay et al, 1972], patas and vervet monkeys [Clarke et al, this volume], and orangutans and chimpanzees [Maple and Zucker, 1978]. Additionally, Bernstein and Gordon [1980] have reported interactions among hybrid macaques, as well as a series of introductions utilizing groups of different species (e.g., vervets and bonnet monkeys, vervets and mangabeys).

The distinction has been made between interspecific and polyspecific interactions [Gartlan and Struhsaker, 1972; Rose, 1977]. The term "polyspecific interaction" indicates that two or more species are "in association," which, in turn, is defined in terms of spatial relationships between the species, and is devoid of any other social or ecological implications [Gartlan and Struhsaker, 1972]. In contrast, "interspecific interaction" is reserved for those instances where actual social interactions between species members occur. The questions of whether or not polyspecific associations occur more or less frequently than expected by chance, and have durations shorter or longer than expected by chance, have been addressed by Waser [1982, 1984]. As Waser [1984] pointed out, the behavior that occurs between species may have adaptive value to the participants, and is independent of whether or not the members of the two (or more) species were in proximity due to chance, or whether there was some social or ecological basis for the association. Thus, a polyspecific interaction is a necessary antecedent for interspecific interactions.

With respect to an ordering of the types of multispecies interactions in terms of frequencies or probabilities of occurrence, Rose [1977] suggested that beyond polyspecific aggregations, simple avoidance was the next most common, and for direct interaction (true interspecific interactions), aggression was the most common form. Nonaggressive interactions, when they do occur, typically involve grooming and play [Rose, 1977]. These latter two behavioral categories are usually interpreted as being affiliative, and conceptually, are probably a step nearer to the social integration of members of different species. Examples of interspecific social integration under natural conditions include a male gibbon (*Hylobates lar*) travelling, feeding, fleeing, resting, and sleeping amidst a group of banded leaf monkeys (*Presbytis melalophus*) [Bernstein, 1967] and "adoption" of two rhesus monkey juveniles by hanuman langurs (*P. entellus*) [Singh and Sen, 1977–78]. The rhesus were observed playing, grooming, and feeding with the langurs. In these cases, individuals of one species became members of social groups of another species.

In this chapter, the polyspecific and interspecific interactions between free-ranging patas monkeys and rhesus monkeys sharing an island habitat are described quantitatively. As the social organization of patas monkeys was the focus of the study [Kaplan and Zucker, 1980; Zucker and Kaplan, 1981a, 1981b], and as the bahavior of African monkeys in comparative perspective is the focus of this volume, the analyses and descriptions included here take a patas perspective. That is, the identities of the patas participants are included (when known), while rhesus are identified only by age and/or sex classes.

METHODS

Subjects and Study Site

At the time of this study, the 36-hectare island of El Guayacan, located off of the coast of southwestern Puerto Rico near the town of La Parguera [see Vandenbergh, 1967; Post and Wiley, 1976; Kerber et al, 1979], supported two species of free-ranging monkeys: a 56-member group of patas monkeys (*Erythrocebus patas*); and two groups of rhesus monkeys (*Macaca mulatta*). These two groups of rhesus together numbered approximately 100 individuals [Herbert, personal communication]. In addition to the rhesus groups, there were several nongroup adult male rhesus monkeys. Between the first and second periods of study (see Data Collection, below), one of the nongroup males joined the smaller of the two rhesus groups. Also located on the island were four 1/4-acre corrals that housed additional rhesus. These corral-housed rhesus were not included in the estimate of the rhesus population, nor did they enter into any of the interactions described in this chapter. Only the interactions of free-ranging monkeys are included here.

None of the free-ranging rhesus monkeys were purposely released on El Guayacan; they had "migrated" (presumably by swimming) from the neighboring island of La Cueva. This migration occurred following the addition of 375 females to the existing La Cueva population [see Kerber et al, 1979; Berkson, 1977]. The two islands are separated by approximately 305 m (1,000 ft) of shallow water, interspersed with several small mangrove islands [see Fig. 1 in Kerber et al, 1979]. The rhesus might also have migrated by travelling through the mangrove swamps, traversing the distance between the islands on the Puerto Rican mainland. Whereas the mainland was easily reached from La Cueva, El Guayacan was more difficult to reach without swimming some of the distance.

The patas group had been released on El Guayacan in 1974, and was comprised of wild-born individuals that had been housed for some period of time under laboratory conditions, first in individual cages, and later, in small social groups on La Cueva. The original stock of free-ranging patas consisted of 27 individuals. At the time of study (1977–1978), the patas group had not been regularly trapped, censused, or tattooed since its original release.

At the start of this study (June, 1977), the group contained one adult male, 18 adult females, 14 dependent infants, and 22 additional immature monkeys. The first infant of the 1978 birth season was born on the last day of the study, bringing the total number of monkeys in the group to 56. One nongroup solitary male patas also lived on the island, and his interactions with rhesus also are included.

The monkeys on El Guayacan were provisioned with commercial monkey diet, which was available ad libitum from two aluminum food hoppers located approximately 250 m apart. Water was also available from two founts located in the vicinity of the food hoppers. The source of water was both collected rainwater and water brought to the island from the Puerto Rico mainland. Unlike the Caribbean Primate Center's Cayo Santiago facility, the two La Parguera islands received comparatively little rainfall [Varley and Vessey, 1977].

Data Collection

Data were collected by ad libitum sampling during two 2-month periods (June 19, 1977 through August 14, 1977; December 9, 1977 through February 5, 1978). These periods correspond to portions of the birth and breeding seasons, respectively. A total of 395.9 hours of observations were made on the patas monkeys, nearly evenly distributed between the two periods (198.3 and 197.6). Notes were taken in diary fashion; they described the observed interactions between individual patas, and between patas and rhesus. Adult patas were individually identifiable by physical characteristics and behavior. Immature patas were identified according to age; sex could not be determined by observation alone [see Hall, 1965]. Rhesus were identified according to age class, and if adult, according to sex. Ad libitum notes were supplemented by scan samples at predetermined intervals. During each scan, the location of each visible patas monkey was noted, as were the identities of the adults. These scans were used to determine individuals' and age classes' relative observabilities [see Kaplan and Zucker, 1980].

Data collection began between 0630 and 0645 hr, at which time there was sufficient light to be able to see any identifying marks on the monkeys. Observations continued until 1800 hr during the first 2-month period, but ended at 1400 hr during the second study period.

The patas monkeys tended to occupy the central portions of the island, whereas the rhesus tended to travel through, and occupy, the mangrove trees that ringed the island [Kerber et al, 1979], although this was not an exclusive arrangement. For example, the patas group ventured into the mangroves on a close to daily basis, usually for 2–3 hr per day. When the mangroves located to the south of El Guayacan proper were entered, tracking and observing the group was not possible. The extensive mangrove swamp to the south is depicted in Vandenbergh [1967] and Kerber et al [1979].

The majority of interactions between the species occurred in the vicinities of the two feeding stations. These stations were situated close to the edges of the island near the mangroves, which aided in transporting monkey diet and water.

Data Analysis

Interactions between species were extracted from the handwritten notes for further analysis. The patas participants were identified by name if adults, or by age class only if immatures. The age classes of the rhesus participants were also noted, as was the sex of the rhesus participant if an adult. If the age class of a participant could not be determined, then it was classified as "unknown," and tabulated as a separate category. Similarly, if three or more individuals of one species were involved, the interaction was classified as involving "multiple animals," and was tabulated as a separate category.

Each interaction was characterized as either polyspecific or interspecific. Polyspecific interactions included proximity, parallel travel, and simultaneous presence and/or feeding at a food hopper. Interspecific interactions were those that involved direct interaction, such as agonism (including spatial displacements) and affiliation (including groom, play, and alloparental behavior). The initiators of these activities were noted. The specific behaviors used in dyadic agonistic interactions between adults were also extracted from notes, and analyzed for both severity and species differences, particularly those behaviors that initiated the agonistic interactions.

With respect to the severity of agonism, agonistic behaviors that involved contact were considered more severe than those that did not involve contact. Similarly, those behaviors that involved gross motor movements and/or movement through space (e.g., charge, chase) were considered more severe than those that involved fine motor movements and/or stationary postures (e.g., vocal threat, head bob, brow flash). Species differences were assessed by converting the frequencies of the various behaviors (or groups of behaviors) to proportions, and then testing for differences between the proportions, as described by Bruning and Kintz [1968]. For these statistical tests, the null hypothesis is that the two proportions are equal. Differences between the behaviors used by patas females in their interspecific agonistic interactions and their intraspecific agonistic interactions [Kaplan and Zucker, 1980; Zucker, unpublished manuscript] were also assessed.

The adult female patas monkeys could be ranked hierarchically based on the outcomes of intraspecific agonistic interactions [Kaplan and Zucker, 1980; Zucker, this volume]. These females' patterns of interactions with rhesus were tested for correlations based on rank, as well as for differences associated with relative dominance status. Data for individual adult patas were corrected for differential observability by dividing the frequency of the behavior being analyzed by the proportion of scan samples in which that individual was observed [see Drickamer, 1974]. There was no relationship between the females' agonistic dominance ranks and observability ($r = .23$, $df = 16$, $p > .05$), nor were there differences based on relative ranks (high, middle, and low; $n = 6$ in each group; $F = 1.71$, $df = 2,15$, $p > .05$).

RESULTS

For all age and sex classes combined, 1,409 interactions (3.5/hr of observation) between species were observed. Of this total, 866 (61.5%) were polyspecific, and 543 (38.5%) were truly interspecific, involving some type of agonistic behavior. Overall, patas initiated 149 (27.4%) of the agonistic interactions. When only dyadic interactions are considered, the percentage of patas-initiated interactions increased slightly, to 28% of the total (137/489). No true affiliative behaviors, such as grooming or play, were observed to occur between the two species.

Polyspecific Interactions

The frequencies of polyspecific interactions, by age and sex classes, are presented in Table I in a matrix format. For these types of interactions, no initiator could be reliably determined. For the group-living patas male, just over half of his polyspecific involvements (54%) were with adult male rhesus, and 32% involved adult female rhesus (see Fig. 1). Only 5.5% of the male's polyspecific interactions involved immature (noninfant) rhesus. Although the actual number of adult rhesus is unknown, there were relatively fewer adult male rhesus than adult female rhesus, which magnifies the mean number of interactions per individual adult rhesus. The nongroup patas male

TABLE I. Matrix of Polyspecific Interactions by Age and/or Sex Class

	Rhesus					
Patas	Adult males	Adult females	Immatures	Unidentified	Multiple	Total
Group male	39	23	4	1	5	72
Adult females	296	120	102	3	42	563
Immatures	134	34	41	1	11	221
Nongroup male	0	1	0	1	0	2
Multiple	6	0	0	0	2	8
Total	475	178	147	6	60	866

Fig. 1. An example of a polyspecific association. The group-living adult male patas monkey feeds while an adult female rhesus (with an infant) sits on the opposite side of the food hopper. A patas female with an infant sits nearby, and an immature rhesus monkey on the far side of the hopper looks at the male patas. Note also the rhesus monkey sitting in the "X" of the poles. (Photo by E. L. Zucker, courtesy of the Caribbean Primate Research Center)

rarely was observed interacting with rhesus, and only once did his polyspecific involvement include an adult rhesus monkey.

Adult female patas monkeys were involved in polyspecific interactions on 563 occasions (Table I), which was a mean of 31.3 interactions per female. For individual females, the number of polyspecific interactions with adult rhesus ranged from nine to 60 ($\bar{X}$ = 22.9, SD = 13.3; Table II). Interactions with adult male rhesus by adult female patas were nearly 2.5 times more

TABLE II. Polyspecific Interactions of Adult Female Patas Monkeys With Adult Rhesus Monkeys

Name[a]	Total number of interactions	Number with rhesus		Observability[b]
		Males	Females	
BE	12	8	4	105
SA	11	7	4	121
TR	16	12	4	180
CI	13	10	3	99
VE	29	19	10	109
LE	9	6	3	125
YI	19	12	7	181
PA	27	19	8	66
AA	29	21	8	98
W	23	18	5	98
SP	15	9	6	161
LO	35	27	8	188
DO	27	24	3	98
SR	14	11	3	68
MA	14	11	3	30
PE	60	36	24	110
GA	45	33	12	150
PO	15	10	5	85
Total	416	296	120	

[a]Two-letter codes are used for names (except for "W"). Females are arranged according to dominance ranks, with "BE" the highest-ranking [see Kaplan and Zucker, 1980].
[b]From Kaplan and Zucker [1980]. This is the number of appearances in the 870 scan samples.

prevalent than were interactions between adult females of the two species. There were 16.4 interactions per patas female with rhesus males, and 6.7 per patas female with rhesus females. The adult patas female could not be positively identified in three interactions with rhesus males.

The agonistic rank of the female patas was significantly related to the frequency of polyspecific interactions with adult male rhesus ($r = .58$, $df = 16$, $p < .05$; frequencies adjusted for observabilities). Lower-ranking females associated more with rhesus males than did higher-ranking females. (This correlation is positive because lower ranks have higher ordinal numbers.) There were significant differences in these adjusted frequencies among high, middle, and low ranking females ($F = 5.51$, $df = 2{,}15$, $p < .05$), with the greatest difference being between the highest and lowest ranking thirds of the hierarchy (Newman-Keuls test). Polyspecific interactions with adult female rhesus were not related to the ranks of the female

Fig. 2. Two immature rhesus monkeys play near the food hopper, while an immature patas sits in the bushes just to the left of the rhesus. (Photo by E. L. Zucker, courtesy of the Caribbean Primate Research Center)

TABLE III. Interspecific Agonistic Interactions by Age and/or Sex Class

	Rhesus					
Patas	Adult males	Adult females	Immatures	Unidentified	Multiple	Total
Group male	19	15	2	2	5	43
Adult females	127	127	77	12	21	364
Immatures	41	38	27	2	4	112
Nongroup male	0	0	0	0	0	0
Multiple	9	3	2	0	10	24
Total	196	183	108	16	40	543

patas ($r = .43$, $p > .05$), nor were there differences due to the relative status of the patas females ($F = 1.67$, $df = 2,15$, $p > .05$).

Like the adult patas, immature patas were in polyspecific associations with male rhesus more than any other age-sex class. Unlike the adult patas, however, their next most frequent association was with immature rhesus (see Table I and Fig. 2).

Agonistic Interactions

The frequencies of agonistic interactions, irrespective of initiators, that occurred between the two species, are presented in Table III in a matrix format by age and sex classes. Agonistic interactions with adult rhesus accounted for between 70 and 79% of the total for the three patas age/sex categories: 79% for the adult male, 70% for the adult females, and 71% for the immatures. The nongroup patas male was not observed interacting agonistically with any rhesus.

The numbers, and corresponding percentages, of agonistic interactions initiated by patas monkeys against rhesus monkeys are presented in Table IV. Of particular noteworthiness is: 1) the higher percentage of initiations by the patas male against rhesus males, 2) the higher percentage of initiations by the patas females against immature rhesus, and 3) the higher percentage of initiations by immature patas against immature rhesus (relative to adult rhesus). Overall, the adult male patas had the highest percentage of initiated agonism (33%).

The raw frequencies of agonistic interactions for the individual adult females with adult rhesus, as well as the frequencies corrected for observability, are presented in Table V. These data are also partitioned according to the sex of the adult rhesus participant. When observability is taken into account, the females' ranks were significantly correlated with frequencies of interactions with adult male rhesus ($r = .57$, $df = 16$, $p < .05$) and adult female rhesus ($r = .61$, $df = 16$, $p < .05$). These correlations indicate that frequencies of interactions with adult rhesus of either sex increased as dominance rank decreased.

When the data in Table V were tested for differences among high, middle, and low ranking segments of the group (e.g., thirds), the aforementioned correlations were borne out statistically. In interactions with adult rhesus males, there were significant differences among the segments of the hierarchy ($F = 4.36$, $df = 2,15$, $p = .03$), with the largest difference between the upper and lower thirds of the hierarchy (Newman-Keuls test). Patas females' agonistic interactions with adult rhesus females also showed differences based on their relative status within the patas group ($F = 7.18$, $df = 2,15$, $p = .006$). In this latter situation, the lower ranking females were significantly more involved agonistically than either the high or middle ranking patas females (Newman-Keuls test).

Severity of Interspecific Agonism

There were 227 agonistic behaviors observed in the interspecific agonism between two adults. Rhesus monkeys emitted 75% of these behaviors. For

TABLE IV. Number and Percentage of Agonistic Interactions Initiated by Patas Monkeys

	Rhesus						
Patas	Adult males (%)	Adult females (%)	Immatures (%)	Unidentified (%)	Multiple (%)	Total	Weighted (%)
Group male	8 (42)	4 (27)	1 (50)	1 (50)	0 (0)	14	33
Adult females	28 (22)	26 (20)	45 (58)	5 (42)	5 (24)	109	30
Immatures	6 (15)	5 (13)	8 (30)	0 (0)	1 (25)	20	18
Multiple	2 (22)	0 (0)	0 (0)	0 (0)	4 (40)	6	25
Total	44 (22)	35 (19)	54 (50)	6 (38)	10 (25)	149	27

TABLE V. Agonistic Interactions of Adult Female Patas Monkeys With Adult Rhesus Monkeys

Name	Total number of interactions	With rhesus males		With rhesus females	
		Freq.	Corrected[a]	Freq.	Corrected
BE	3	2	.0190	1	.0095
SA	4	1	.0083	3	.0248
TR	4	3	.0167	1	.0056
CI	6	4	.0404	2	.0202
VE	14	10	.0917	4	.0367
LE	0	0	.0000	0	.0000
YI	19	8	.0442	11	.0608
PA	15	8	.1212	7	.1061
AA	14	9	.0918	5	.0510
W	8	5	.0510	3	.0316
SP	9	8	.0497	1	.0062
LO	16	11	.0585	5	.0266
DO	21	8	.0816	13	.1327
SR	9	4	.0588	5	.0735
MA	7	4	.1333	3	.1000
PE	58	23	.2091	35	.3182
GA	34	11	.0733	23	.1533
PO	10	5	.0588	5	.0588
Total	251	124		127	

[a]Corrected frequencies are raw frequencies divided by that individual's observability score (from Table II).

both species, however, noncontact agonism predominated, and only one contact behavior was observed for each species; a patas female hit a rhesus female, and a rhesus male bit a patas female. Agonism directed toward rhesus adults by patas adults was 98% noncontact, and patas-directed agonism by rhesus was 99% noncontact.

Patas initiations. With respect to the behaviors used in initiating agonistic interspecific interactions, adult patas females used threats predominantly (head bob, browflash, vocalizations, yawns), regardless of the sex of the adult rhesus recipient. Patas females were not observed to spatially displace any adult rhesus, although this was the single most common form of patas female-female (intraspecific) interaction. Patas females, however, threatened rhesus females significantly more, proportionately, than they did patas females ($z = 17.2$, $p < .001$). Patas females' interspecific interactions with rhesus females also included proportionately more chasing ($z = 4.1$, $p < .01$) and charging and lunging ($z = 2.5$, $p < .01$) than their intraspecific interactions.

Patas females, in initiating agonism with rhesus males, were most likely to use a threat of some kind ($p = .89$). Patas females threatened rhesus males significantly more, proportionately, than they received threats from rhesus males ($z = 17.4$, $p < .001$).

Few initiating behaviors of the adult male in dyadic contexts were observed. However, the group-living patas male, like the patas females, used threats predominantly in agonistic interactions with adult rhesus females (five out of the six behaviors observed). The remaining behavior was a displacement of an adult female rhesus. No initiating behaviors between the patas male and adult rhesus males were observed.

Rhesus initiations. Agonistic interactions initiated by rhesus females, and involving patas females, were most likely to begin with a spatial displacement of a patas female ($p = .72$). Threats and charges/lunges were the next most likely behaviors to have occurred (.13 and .12, respectively). Similarly, rhesus males' agonistic interactions with patas females were most likely to have begun with a spatial displacement ($p = .73$; Fig. 3a–c). There were no significant sex differences for rhesus' behaviors that initiated agonistic interactions with patas females. When the patas male was the recipient of a rhesus male's agonism, displacements were the most common agonistic behavior ($p = .44$), although the total frequency for male-male interactions was low ($n = 9$). Threats by rhesus males occurred 3 times, and charges and lunges occurred twice.

Species differences could be assessed using the behaviors of the adult females of both species, as these behaviors occurred with sufficient frequen-

Fig. 3. A displacement of three adult female patas by a single adult male rhesus. A) Three adult female patas sit in proximity in the shade of the roof of an old feeding station. B) The three female patas move away from their prior positions. C) The females go to the edges of the cement block as a male rhesus sits in the corner where the females had been. (Photos by E. L. Zucker, courtesy of the Caribbean Primate Research Center)

cies to allow statistical comparisons. Proportionately, rhesus females displaced patas females significantly more than the reverse ($z = 15.5$, $p < .001$; patas females did not displace rhesus females), whereas patas females engaged in proportionately more threats ($z = -7.0$, $p < .001$) and chases ($z = -2.2$, $p < .05$). There were no species differences in the proportions of charges and lunges.

Agonistic aiding. Patas monkeys were observed entering an ongoing interspecific agonistic interaction [termed "agonistic aiding" by Massey (1977) and "fight interference" by Kaplan (1977)] on 18 occasions. Of these 18, eight (44%) were by the adult male of the patas group, and nine (50%) were by adult female patas. Two thirds of the agonistic aids (12/18) occurred during agonistic interactions with adult rhesus monkeys.

DISCUSSION

The interactions between the patas and rhesus monkeys on the island of El Guayacan were predominantly polyspecific. Interactions that were truly interspecific were exclusively agonistic, as affiliative behaviors between members of the two species were not observed.

Overlap of Resources and Home Range

The concentration of food sources, brought about by the need for provisioning, likely increased the probability that the two species would encounter each other as often as they did (3.5 associations or interactions per hour). Polyspecific and interspecific interactions, however, did occur elsewhere on the island, and were presumably independent of feeding and drinking. In addition to the concentration of food, the locations of the feeders probably played a facilitating role. The feeders were located at the intersections of the rhesus-preferred mangrove ring surrounding the island and the patas-preferred island proper. Whereas these locations were selected for ease of delivering food and water to the island, the location likely had an effect on the behavior of the monkeys. These factors, a common and concentrated food source and range overlap, are frequently cited as the ecological variables contributing to polyspecific interactions among populations of feral monkeys [i.e., Gartlan and Struhsaker, 1972; Struhsaker and Leyland, 1979; Struhsaker, 1981; Rose, 1977; Waser, 1980, 1982, 1984].

Although food and range overlap might increase the likelihood of polyspecific associations, even if these associations are due to chance [e.g., Waser, 1982], these conditions do not account for all polyspecific associations [Struhsaker, 1981], nor are they predictive of the occurrence of true interspecific interactions. For example, Moreno-Black and Maples [1977] described a site in Kenya where four species were found in a limited area (approximately .25 km^2; density estimated at 252/km^2), but the habitat was apparently diverse enough in terms of food types and availability that little resource overlap was found. Interspecific interactions between any two of the four species (vervets, Sykes' monkeys, Angolan colobus, and yellow baboons) were relatively few in number.

An additional factor that should be taken into account when considering the contributuions of resource overlap versus separation, particularly as applied to the patas and rhesus described here, is the degree of terrestriality. Patas are the most terrestrial of the *Cercopithecus-Erythrocebus-Miopithecus* group [Hall, 1965; see also Clarke et al, this volume], and Lindburg [1971] described the daily foraging pattern of Indian rhesus as similar to patas. Lindburg did not directly address the question of terrestriality, but mentioned that forest groups of rhesus spent the nights in the trees wherever they had been foraging at the end of the day (p. 24). From these characterizations of the tendencies for patas and rhesus groups to spend a large proportion of time on the ground, the fairly high frequency of proximity between members of these two species is further explained. Whereas there may be concentrated food, intersections of ranges, and high social densities, such as described by Moreno-Black and Maples [1977], the degree of vertical partitioning of space appears to be a critical variable. Had one of the species on El Guayacan been a species that was more arboreal than either patas or rhesus, it might be expected that there would have been less association between the two species.

Demographics of Polyspecific Associations

Patas of each age/sex class delineated were most frequently in polyspecific associations with adult male rhesus. These rhesus males were typically those that lived independently of the two heterosexual groups on El Guayacan. During the first study period, these males lived as a three-member "group," and subsequently, as a two-member group following the integration of one of these males as the top-ranking male in the smaller of the two heterosexual groups. If one of these extra-group males was in the vicinity of a food hopper, then any noninfant patas monkey would continue its approach to that feeder, although typically the patas would feed from the opposite side of the food hopper from the rhesus (see Fig. 1). Conversely, if patas were in the vicinity of a food hopper, and an extra-group rhesus male approached, then the patas would remain where they were (see Interspecific Agonism below, regarding spatial displacements).

An alternative explanation for the higher number of observed polyspecific associations with adult rhesus males relative to adult females is that rhesus males were more observable than were females. The observabilities of the patas are known [Kaplan and Zucker, 1980], but the differential observability of rhesus age/sex classes was not determined as part of this study. However, observability data from Kaplan's [1977] study of fight interference among free-ranging rhesus monkeys on Cayo Santiago provides evidence contrary to this alternative explanation. For Group F on Cayo Santiago, adult rhesus females were significantly more visible than were adult males (natal plus non-natal) [data from Kaplan, 1977, Table 1]. There was also a significant difference, however, between the natal and non-natal males of that group with respect to observability. If the less observable non-natal males are excluded from the comparison with adult females, then there is no longer a significant difference between the two sexes of adults (t-tests, $\alpha = .05$). Regardless of the groups of adult rhesus included in these comparisons, the observability of adult males as an age/sex class did not exceed the observability of adult females.

Adult female patas' polyspecific involvements with adult rhesus varied according to the sex of the adult rhesus and the social status of the female patas. As discussed above, there were more polyspecific associations with male rhesus than with female rhesus. In addition, lower ranking female patas were more likely to be in polyspecific associations with male rhesus than were higher ranking females. This difference cannot be due to observability differences, as, unlike adult female rhesus [data from Kaplan, 1977], there was no significant relationship between observability and agonistic rank for adult female patas monkeys.

As most of the polyspecific associations occurred in the vicinities of the two food hoppers, it is possible that the lower ranking females spent more time near the food hoppers, or that they were more likely to approach the hoppers in advance of higher ranking females. Data are available to test the latter hypothesis, but are only partially supportive of this idea. Middle and low ranking females approached the food hoppers and ate first more often than was expected by chance during the morning feeding periods, while the high ranking females were first-to-feed less often than expected [Zucker, unpublished]. Overall, however, the number of times an adult female was the first one to approach a feeder was not significantly related to dominance rank.

Alternatively, the lower ranking females might have been the last ones to eat, after the rest of the patas group had moved away from the feeding areas, and thus remained in the feeding areas when rhesus moved into the areas.

Again, the propensity for middle and low ranking females to eat first more often than was expected indirectly argues against this "last-to-feed" hypothesis. That these lower ranking females could eat first, ahead of the high ranking females, detracts from a competitive exclusion explanation. Data directly testing this last-to-feed hypothesis, however, are not readily extractable from the field notes. In this type of situation, the males that would be moving into the feeding areas would be the extra-group rhesus males, as the patas would not remain at the hoppers if the rhesus groups approached (see Interspecific Agonism below).

Although the influx of the rhesus groups into the feeding areas usually was sufficient to displace any patas from the hoppers and the immediate surrounding, patas would gradually return either to feed opposite the rhesus or be spatially near to the rhesus (as in Fig. 2), even if the rhesus were adult females. In contrast to the female patas' rank-related patterns of polyspecific associations with male rhesus, similar associations with female rhesus were independent of rank.

Interspecific Agonism

In contrast to the patterns observed for patas' polyspecific associations, where adult male rhesus were more likely to be involved than adult female rhesus, patas' agonistic interactions with adult rhesus monkeys were more evenly distributed across the two sexes. This even distribution applied also to the agonistic interactions of immature patas, and to the numbers of interactions initiated by patas.

Overall, the percentage of interactions initiated by the patas was low (approximately 27%), and initiations by the adult male patas with adult rhesus were consistently higher than the number (or percentage) of initiations by adult female patas with adult rhesus. Adult female patas, however, initiated more than half of the agonistic interactions that they had with immature rhesus. Struhsaker [1981] and Southwick and Southwick [1983] explained outcomes of mixed-species agonism on the basis of the relative body sizes of the species or individuals involved. Struhsaker [1981] reported that redtail monkeys (*Cercopithecus ascanius*) lost approximately 90% of their aggressive interactions with other species, particularly blue monkeys (*C. mitis*). Similarly, Southwick and Southwick [1983] found rhesus monkey males able to dominate the crab-eating macaque male (*Macaca fascicularis*) that was in the group. Although Southwick and Southwick did not have weights or measurements on these males, relative size and dominance status were thought to be positively related. For the present study, the advantage that the adult female patas had over the immature rhesus might be explained by size differences, although this explanation does not hold for the interaction patterns of other age/sex classes. Patas of any age/sex class outsized their rhesus counterparts, yet rhesus tended to dominate. The propensity for rhesus monkeys to dominate other species may be a reflection of their baseline level of aggressiveness, for which they are well known [i.e., Fairbanks et al, 1978; Thierry, 1985; Kaplan, this volume]. Kaplan [this volume] has elaborated on the aggressive styles of rhesus monkeys and other species of Old World monkeys.

Comparisons With Intergroup Agonism

Another way to view and interpret the agonism between two species is to compare the interspecies agonism with the agonism that occurs between two groups of the same species (i.e., intergroup aggression). For patas, descriptions of intergroup aggression are limited to two encounters described by Hall [1965]. One interaction involved the adult male of one group (group IV) initiating a chase of a second group (group III). This male was joined in the chase by other group members. Hall estimated that members of group III ran approximately one mile before stopping. The second episode involved the same two groups, but in this case, members of group III left an area, apparently after the male of group III spotted group IV. In this retreat, the male of group III was the last to leave the area. This male-male intolerance in consistent with the finding of Kummer [1974], who experimentally paired patas males in an outdoor corral, as well as what was seen between the group-living patas on El Guayacan and the extra-group male on that island [Kaplan and Zucker, 1980].

A different view of male-male relationships has been described by Chism and Rowell [1986] and Harding and Olson [1986], with both reports describing multimale membership, and the fluidity of male membership, in two patas groups at the Mutara Ranch in the Laikipia district of Kenya. No details of intergroup interactions were provided, except that ''groups had broadly overlapping home ranges and encountered each other frequently'' [Chism and Rowell, 1986, p. 32].

The style of interaction between patas groups, as described by Hall [1965], is similar to what was observed between the patas and rhesus on El Guayacan. The data regarding the severity of agonistic interactions indicate that adult patas of both sexes were most likely to be displaced by the approaching rhesus adults. When displaced, however, the patas did not retreat as far as the Ugandan patas. This reaction to an interspecies conflict is also consistent with Hall's description of patas-baboon interactions, as well as with patas intragroup conflict [Kaplan and Zucker, 1980]. Patas-vervet interactions are apparently more variable in terms of outcomes [see Clarke et al, this volume].

Greater detail about intergroup aggression is available for feral and free-ranging rhesus monkeys. In India, intergroup avoidance was common, with smaller groups retreating from larger groups, although an exchange of threats accompanied some of these interactions [Southwick, 1962]. Southwick [1962] also described intergroup "fights" as occurring daily, and these initially involved the adult males. All other age/sex classes, including infants, became involved in these interactions. For rhesus on La Cueva island (at La Parguera), outcomes of intergroup displacements were also related to group size, and if direct confrontation were to occur, adult males were the primary participants [Vessey, 1968]. For one of the groups at La Cueva (group C), dominance rank among males was inversely related to frequency of intergroup interaction. Peripheral males were most likely to be involved, both as aggressors and as victims. Hausfater [1972] used the term "partial-group interactions" for rhesus on Cayo Santiago, a term that also describes the interactions between the patas and rhesus on El Guayacan. Hausfater's data (from his Table IV) on frequencies of partial-group aggression indicate that each sex class of monkeys, 2 years of age and older, engaged in an average of at least 18 partial-group aggressive interactions. For such rhesus in group F, there was no significant sex difference in numbers of partial-group aggressive interactions (2-year-olds and above, nonmating season data, t-test, $\alpha = .05$).

These descriptions of rhesus intergroup aggressive interactions are similar to the patterns exhibited by adult female patas in the following ways. First, the inverse relationship between rank and interspecific involvement described for male rhesus by Vessey [1968] also describes the pattern for female patas' interactions with adult rhesus of either sex. Secondly, the rates of aggressive partial-group interactions, calculated from Hausfater's [1972] data, are very similar to those calculated for patas females. From Hausfater's data, rhesus females 2 years of age and above were in .57 partial-group aggressive interactions per hr (27 females, 524 nonmating season interactions, 911 hours of observation). For the patas of the present study, they were involved at an estimated rate of .64 per hour [18 females, 254 interactions with adult rhesus (Table III), 395.9 hours of observation]. Due to the sampling techniques used in Hausfater [1972] and the present study, these rates should be taken as estimates rather than exact calculations. Nonetheless, the similarity remains.

The male patas' involvement in patas-rhesus agonism follows from the patterns of intergroup aggression of both species. The male patas' social "role" is described as being primarily vigilant and spatially peripheral, rather than being an integral member of the social group [Hall, 1965; Struhsaker and Gartlan, 1970]. The patas male likely is most reactive to other males (although perhaps more so to unfamiliar males), as are rhesus males. Male-directed vigilance, and any subsequent actions, are consistent with the findings of the present study, whereby male patas initiated a higher proportion of agonism against adult male rhesus relative to adult female rhesus. As female rhesus outnumbered male rhesus on El Guayacan, this male-male pattern is evident also in the total frequencies of agonistic interactions.

Also consistent with the behavior of feral male patas was the observed pattern of agonistic aiding. This adult male patas was not observed to interfere in intragroup conflicts [Kaplan and Zucker, 1980], but did interfere in extra-group, interspecific conflicts. He was involved in nearly half of all of the observed aiding. This difference in the male's behavior in intragroup versus extra-group conflicts indicates the patas males' extra-group orientation, as has been described by field researchers [e.g., Hall, 1965; Struhsaker and Gartlan, 1970]. This extra-group orientation can now perhaps be extended to include the observation and assessment of males in adjacent groups as a reproductive strategy [e.g., Chism and Rowell, 1986; Harding and Olson, 1986].

Severity of Interspecies Agonism

Adult patas showed a strong tendency to avoid interactions with adult rhesus when rhesus were the initiators, as evidenced by the patas' propensity to be displaced. Although adult patas are larger than adult rhesus, the patas were not successful, conversely, in being able to displace rhesus. In their interactions with rhesus, adult patas used threats (yawns, headbobs, browflashes) most frequently, which is a form of agonism that is more severe than are spatial displacements [e.g., Rose, 1977]. Threats comprised a larger proportion of female patas' repertoires in interspecific interactions than they did in intraspecific conflicts, as did the behaviors of chase, charge, and lunge. Interestingly, rhesus typically use threats more than any other agonistic behavior [Teas et al, 1982]. This pattern shown by the patas indicates that the interspecific interactions were more severe than were patas-patas interactions. More overt forms of communication were needed in the patas' agonistic interactions with rhesus.

The patas' tendency to mediate social conflicts with rhesus, as well as with other patas [Kaplan and Zucker, 1980], by spatial adjustments could stem from the lack of an extensive repertoire of submissive signals [see Kaplan, this volume]. Jacobus and Loy [1981] described three submissive signals for patas, including the grimace and cower that are also shown by rhesus monkeys [Sade, 1973; Redican, 1975; Mitchell, 1979, Plate 22]. The third signal described was a gecker vocalization. Patas, however, lack the communicatory signals of presents and lip-smacks that are found in macaque communication. These differences in communicative signals, coupled with the patas' tendency to be displaced and flee, could have altered the nature of the interactions that occurred. For example, the approaches by rhesus monkeys toward the patas may not have signalled any agonistic encounter (if viewed from the rhesus' perspective), whereas the patas' tendency to retreat defined the interaction as agonistic. It may not be the number or type of dominance/submissive signals that led to interspecific agonism, but the inability for one species to read the "metacommunicative" signals of the other species [e.g., Altmann, 1967]. As rhesus "initiated" about 73% of the agonistic interactions, it was typically the patas who were not "understanding" the rhesus' signals. In other instances, however, unambiguous agonistic behaviors were used by both species. While there are several common behavioral and communicative signals shared by members of these two species, differences remain. These differences likely contributed to the segregration of these species with respect to affiliative interactions.

Species Segregration

Maple [1974a] proposed four factors that, in part, determine the quality and quantity of interactions between species. These factors are proximity, behavioral similarity, factors that force contact ("opportunistic relations"), and behavioral plasticity. Maple [1974b] further hypothesized that if interspecific experience occurred when the subjects were young, then there would be a higher probability of developing viable social relationships. For the monkeys on El Guayacan, several of these factors were operative, such as proximity (same island) and forced contact (common food source). However, the behavioral variables of similarity and plasticity were not as pronounced or prevalent, and their absence likely hindered the occurrence of affiliative social interactions. Kaplan [this volume] elaborated on some of these behavioral differences, such as the use of physical contact, grooming, and fight interference patterns. In Kaplan's analyses, patas and vervets were more similar behaviorally than were patas and rhesus, which suggests that patas-vervet mixed-species groups would show more affiliative interactions than would patas-rhesus. This is what Clarke et al [this volume] found for a mixed-species group of young patas and African green monkeys. Under normal environmental conditions, cross-species affiliative interactions occurred, although preferences for conspecifics were expressed, such as in grooming. In the high density mixed-species group mentioned by Clarke et al, at least one interspecific attachment seemed to have formed [Zucker, personal observation], although this relationship was not permanent, and at times, it resembled a mother-infant relationship (a young male patas ventrally carried a female African green monkey). Under these captive conditions, most of the factors described by Maple [1974a, b] were present. Behavioral plasticity (or adaptability) is more difficult to assess and quantify than are the other variables, although it is generally believed that younger animals are more plastic in their behaviors than are adults (i.e., Struhsaker, 1981; Zucker et al, 1986].

Alternatively, the degree of dietary overlap has been found to correlate with the extent of affiliative social interactions among rain-forest monkeys [Struhsaker, 1981]. Members of species who had little dietary overlap tended to groom each other more than did members of species with greater dietary overlap. This ecological explanation, however, is not mutually exclusive with the behavioral framework proposed by Maple [1974a, b]. Plasticity seems to be able to override food competition, as Struhsaker [1981] found no association between the degree of food overlap and competition, on the one hand, and the tendency for immature monkeys to engage in social play with immatures of other species.

Despite the fact that the patas and rhesus on El Guayacan did engage in interspecific agonism, the majority of interactions were polyspecific, indicative of an overall tolerance of each other. Data regarding the severity of agonism indicated that nearly all of the agonism was of the noncontact type, at least in the agonism that occurred between adults. This high proportion of noncontact agonism further supports that idea of tolerance, although it is not quite the ''pacific'' interactions described by Struhsaker and Gartlan [1970] for patas-vervet relations. Returning to Rose's [1977] ordering of types of interactions, the data for the patas-rhesus interactions clearly fall along the continuum proposed. The most common forms of interactions were polyspecific associations. The next most common was simple avoidance. This was especially the case when adult interactions were analyzed according to severity, as patas adults tended to move away from a rhesus' approach. Among feral populations, aggression was found to precede affiliation, and the data from El Guayacan support this ordering of behaviors. Whereas affiliation and social integration would be expected to follow temporally from aggression, these behaviors and situations were not observed to occur between the patas and rhesus of El Guayacan. At this point, perhaps the behavioral differences began to outweigh the similarities, and further social interaction became impossible. Alternatively, perhaps additional time in a common, yet alien, environment was needed. Developmental changes in the behavior of immatures towards members of the other species might reflect trends towards increased affiliative interactions, perhaps indicating the behavioral plasticity and communicative plasticity that seem to be needed for cross-species social interaction.

CONCLUSIONS

1. Polyspecific and interspecific interactions between patas and rhesus monkeys were observed 1,409 times. These interactions were predominantly polyspecific (61.5%), and the observed direct interactions between species were exclusively agonistic.

2. Most of the adult patas' polyspecific associations were with adult male rhesus. The male rhesus were usually the two or three males that were not associated with either of the two rhesus groups on the island.

3. The social status of adult female patas was related to the frequency of polyspecific associations with adult male rhesus, but was not related to females' associations with adult female rhesus. Lower ranking adult female patas were in polyspecific interactions with male rhesus more frequently than were higher ranking female patas.

4. Agonistic interactions were typically initiated by rhesus monkeys, as patas initiated 27.4% of the total number of interactions. Patas' interspecific agonistic interactions were most frequently with adult rhesus. Lower ranking adult female patas were involved in significantly more interspecific agonism with adult rhesus of both sexes than were higher ranking adult female patas.

5. Agonistic interactions between adults of the two species were characterized by noncontact agonism. The behaviors most likely to initiate agonistic interactions differed for the adults of the two species; rhesus approaches were usually sufficient to spatially displace patas, whereas patas relied on threats as their initiating behavior.

6. Ecological variables, such as a common food source and home range overlap, in addition to the degree of terrestriality, contributed to the high frequencies of polyspecific and interspecific interactions between the two species. The patterns of interspecific agonism observed reflect the manner in which these species interact in conspecific, agonistic encounters. Affiliative interspecific interactions were limited by social and behavioral variables, such as differences in appeasement gestures and, perhaps, metacommunicative signals.

ACKNOWLEDGMENTS

This research was supported, in part, by a faculty grant from the University of Alabama in Birmingham to Jay R. Kaplan, with supplemental funding provided through the Department of Psychology of Emory University. I thank the administration and staff of the Caribbean Primate Research Center at the time of the study, for providing access to the islands, and for their assistance and support. In particular, thanks go to W. T. Kerber, H. J. Herbert, M. J. Kessler, and R. G. Rawlins. On the academic side, the following individuals have had input, over the years, into the ideas and analyses presented here: T. L. Maple, A. T. Pietrewicz, J. R. Kaplan, E. O. Smith, and D. B. Neill. Comments on drafts of this chapter were provided by M. R. Clarke and R. McCleary. R. McCleary also assisted in the tabulation of the "severity of agonism" data. A portion of the data that are included in this chapter was presented orally at the annual meeting of the American Society of Mammalogists, Athens, Georgia, 1978.

REFERENCES

Abordo EJ, Mittermeir RA, Lee J, Mason P (1975): Social grooming between squirrel monkeys and uakaris in a seminatural environment. Primates 16:217–221.

Aldrich-Blake FPG (1968): A fertile hybrid between two *Cercopithecus* spp. in the Budongo Forest, Uganda. Folia Primatol 9:15–21.

Altmann SA (1967): The structure of primate social communication. In Altmann SA (ed): "Social Communication Among Primates." Chicago: University of Chicago Press, pp 325–362.

Baker RP, Preston DG (1973): The effects of interspecies infant interaction upon social behavior of *Macaca irus* and *Erythrocebus patas*. Primates 14:383–392.

Berkson G (1977): The social ecology of defects in primates. In Chevalier-Skolnikoff S, Poirier FE (eds): "Primate Biosocial Development." New York: Garland Publishing, pp 189–204.

Bernstein IS (1967): Intertaxa interactions in a Malayan primate community. Folia Primatol 7:198–207.

Bernstein IS (1968): Social status of two hybrids in a wild troop of *Macaca irus*. Folia Primatol 8:121–131.

Bernstein IS, Gordon TP (1980): Mixed taxa introductions, hybrids, and macaque systematics. In Lindburg DG (ed): "The Macaques." New York: Van Nostrand Reinhold, pp 125–147.

Bernstein IS, Balcaen P, Dresdale L, Gouzoules H, Kavanagh M, Patterson T, Warner P (1976): Differential effects of forest degradation on primate populations. Primates 17:401–411.

Bruning JL, Kintz BL (1968): "Computational Handbook of Statistics." Glenview, IL: Scott, Foresman and Co.

Chism JB, Rowell TE (1986): Mating and residence patterns of male patas monkeys. Ethology 72:31–39.

Drickamer LC (1974): Social rank, observability, and sexual behavior of rhesus monkeys (*Macaca mulatta*). J Reprod Fertil 37:117–120.

Fairbanks LA, McGuire MT, Kerber W (1978): Effects of group size, composition, introduction techniques and cage apparatus on aggression during group formation in rhesus monkeys. Psychol Rep 42:327–333.

Gartlan JS, Struhsaker TT (1972): Polyspecific associations and niche separation of rain-forest anthropoids in Cameroon, West Africa. J Zool 168:221–266.

Hall KRL (1965): Behaviour and ecology of the wild patas monkey, *Erythrocebus patas*, in Uganda. J Zool 148:15–87.

Harding RSO, Olson DK (1986): Patterns of mating among male patas monkeys (*Erythrocebus patas*) in Kenya. Am J Primatol 11:343–358.

Hausfater G (1972): Intergroup behavior of free-ranging rhesus monkeys (*Macaca mulatta*). Folia Primatol 18:78–107.

Hayashi K (1975): Interspecific interaction of the primate groups in Kibale Forest, Uganda. Primates 16:269–283.

Izawa K (1975): Foods and feeding behavior of monkeys in the upper Amazon Basin. Primates 16:295–316.

Jacobus S, Loy J (1981): The grimace and gecker: A submissive display among patas monkeys. Primates 22:393–398.

Kaplan JR (1977): Patterns of fight interference in free-ranging rhesus monkeys. Am J Phys Anthropol 47:279–288.

Kaplan JR, Zucker E (1980): Social organization in a group of free-ranging patas monkeys. Folia Primatol 34:196–213.

Kerber WT, Herbert HJ, Vickers JH (1979): Establishing a free-ranging breeding colony of rhesus monkeys, Part I. J Med Primatol 8:129–142.

Klein LL, Klein DJ (1973): Observations on two types of neotropical primate intertaxa associations. Am J Phys Anthropol 38:649–643.

Kummer H (1974): Distribution of interindividual distances in patas monkeys and gelada baboons. Folia Primatol 21:153–160.

Lindburg DG (1971): The rhesus monkey in North India: An ecological and behavioral study. In Rosenblum LA (ed): "Primate Behavior: Developments in Field and Laboratory Research," Vol. 2. New York: Academic Press, pp 1–106.

Maple TL (1974a): Basic studies of interspecies attachment behavior. Ph. D. Dissertation, University of California, Davis.

Maple TL (1974b): On the need for investigations of inter-species social behavior within the order of Primates. J Behav Sci 2:63–66.

Maple T, Lawson R (1975): Interspecies vocal communication between a baboon and a macaque. Primates 16:99–101.

Maple T, Westlund B (1975): The integration of social interactions between cebus and spider monkeys in captivity. Appl Anim Ethol 1:305–308.

Maple TL, Zucker EL (1978): Ethological studies of play behavior in captive great apes. In Smith EO (ed): "Social Play in Primates." New York: Academic Press, pp 113–142.

Maple T, Scott S, Redican WK (1976): Intergeneric sexual behavior between an adult baboon and an adult macaque. Boll Zool 43:213–220.

Massey A (1977): Agonistic aids and kinship in a group of pigtail macaques. Behav Ecol Sociobiol 2:31–40.

Mitchell G (1979): "Behavioral Sex Differences in Nonhuman Primates." New York: Van Nostrand Reinhold.

Moreno-Black G, Maples WR (1977): Differential habitat utilization of four Cercopithecidae in a Kenyan forest. Folia Primatol 27:85–107.

Nagel U (1973): A comparison of anubis baboons, hamadryas baboons and their hybrids at a species border in Ethiopia. Folia Primatol 19:104–165.

Post W, Wiley JW (1976): The yellow-shouldered blackbird—present and future. Am Birds 30:13–20.

Redican WK (1975): Facial expressions in nonhuman primates. In Rosenblum LA (ed): "Primate Behavior: Developments in Field and Laboratory Research," Vol. 4. New York: Academic Press, pp 103–194.

Rose MD (1977): Interspecific play between free-ranging guerezas (*Colobus guereza*) and vervet monkeys (*Cercopithecus aethiops*). Primates 18:957–964.

Sade DS (1973): An ethogram for rhesus monkeys I. Antithetical contrasts in posture and movement. Am J Phys Anthropol 38:537–542.

Samuels A, Altmann J (1986): Immigration of a *Papio anubis* male into a group of *Papio cynocephalus* baboons and evidence for an anubis-cynocephalus hydrid zone in Amboseli, Kenya. Int J Primatol 7:131–138.

Seay B, Schlottman RS, Gandolfo R (1972): Early social interaction in two monkey species. J Gen Psychol 87:37–43.

Singh RK, Sen NN (1977–78): A note on adoption of two rhesus juveniles by a langur group. J Sci Res: Banaras Hindu University 28:135–138.

Southwick CH (1962): Patterns of intergroup social behavior in primates, with special reference to rhesus and howling monkeys. Ann NY Acad Sci 102:436–454.

Southwick CH, Southwick KL (1983): Polyspecific groups of macaques on the Kowloon Peninsula, New Territories, Hong Kong. Am J Primatol 5:17–24.

Struhsaker TT (1981): Polyspecific associations among tropical rainforest primates. Z Tierpsychol 57:268–304.

Struhsaker TT, Gartlan JS (1970): Observations on the behaviour and ecology of the patas monkey (*Erythrocebus patas*) in the Waza Reserve, Cameroon. J Zool 161:49–63.

Struhsaker TT, Leyland L (1979): Socioecology of five sympatric monkey species in the Kibale Forest, Uganda. Adv Study Behav 9:159–228.

Teas J, Feldman HA, Richie TL, Taylor HG, Southwick CH (1982): Aggressive behavior in the free-ranging rhesus monkeys of Kathmandu, Nepal. Aggressive Behav 8:63–77.

Thierry B (1985): Patterns of agonistic interactions in three species of macaque (*Macaca mulatta, M. fascicularis, M. tonkeana*). Aggressive Behav 11:223–233.

Vandenbergh JG (1967): The development of social structure in free-ranging rhesus monkeys. Behaviour 29:174–194.

Varley MA, Vessey SH (1977): Effects of geographic transfer on the timing of seasonal breeding of rhesus monkeys. Folia Primatol 28:52–59.

Vessey SH (1968): Interactions between free-ranging groups of rhesus monkeys. Folia Primatol 8:228–239.

Waser PM (1980): Polyspecific associations of *Cercocebus albigena*: Geographic variation and ecological correlates. Folia Primatol 33:57–76.

Waser PM (1982): Primate polyspecific associations: Do they occur by chance? Anim Behav 30:1–8.

Waser PM (1984): "Chance" and mixed-species associations. Behav Ecol Sociobiol 15:197–202.

Zucker EL: Severity of agonism in a group of free-ranging patas monkeys. Unpublished manuscript.

Zucker EL, Kaplan JR (1981a): Allomaternal behavior in a group of free-ranging patas monkeys. Am J Primatol 1:57–64.

Zucker EL, Kaplan JR (1981b): A reinterpretation of "sexual harassment" in patas monkeys. Anim Behav 29:957–958.

Zucker EL, Dennon MB, Puleo SG, Maple TL (1986): Play profiles of captive adult orangutans: A developmental perspective. Dev Psychobiol 19:315–326.

Comparative Behavior of African Monkeys, pages 127–150

7

Dominance and Affiliation in the Cercopithecini and Papionini: A Comparative Examination

Jay R. Kaplan

Department of Comparative Medicine, Bowman Gray School of Medicine of Wake Forest University, Winston-Salem, North Carolina 27103

INTRODUCTION

The nonleafeating Old World monkeys are sometimes divided on the basis of morphological and chromosomal characteristics into two tribes, the Cercopithecini and the Papionini [Delson and Andrews, 1975]. The tribe Cercopithecini consists of the guenons (*Cercopithecus*) and three closely related genera (*Erythrocebus, Allenopithecus,* and *Miopithecus* [Delson and Andrews, 1975; Jolly (1966) did not include *Miopithecus* as a member of the tribe]. The tribe Papionini consists of *Papio* (sensu lato), *Macaca, Cercocebus,* and *Theropithecus* [Delson and Andrews, 1975]. While the two tribes have been compared morphologically and genetically [e.g. Giusto and Margulis 1981; Dutrillaux et al 1982], there has been no systematic comparison of social behavior. In fact, the majority of studies of primate social behavior have been done on Papionini species [Erwin and Zucker, this volume]; as a result, certain aspects of their social behavior, for example social dominance patterns, have sometimes been taken as representative of all the nonleafeating Old World monkeys [Wilson, 1975; Chalmers, 1980].

Behavioral generalizations across these tribes may not be valid, as evidence exists indicating that the social behavior of at least some members of the Cercopithecini differs markedly from that of the Papionini. Hall [1965], for example, commented extensively on the propensity of patas monkeys (*Erythrocebus patas*) to utilize a social and ecological strategy of avoidance and concealment in contrast to the baboon tendency for conflict and confrontation. Other patas characteristics noted by Hall included the absence of appeasement gestures and the social remoteness of the adult males. Hall attributed these characteristics of the patas monkeys to their extreme adaptation to a life on the ground.

While guenons are not nearly as terrestrial as patas monkeys, similar behavioral features appear to differentiate both groups from the baboons. This was shown by Rowell [1971], who contrasted the agonistic behavior of *Cercopithecus mitis* and *C. aethiops* with that of *Papio anubis*. She suggested that among the notable characteristics of these guenons were: 1) the presence of adult females rather than males as the most dominant and socially active animals in the group; 2) the relative absence of gestures of social subordination or appeasement; 3) the absence of fight interference or policing behavior; and 4) the increased reliance on physical contact during fights. She identified these characteristics and the long, sharp canines of females as instrumental in leading to the observation of several fatal fights (''murders'') among captive Sykes' and vervet monkeys. Fatal encounters were not observed among similarly housed baboons.

In this chapter, dominance and affiliation data collected by the author and his colleagues on patas, vervet, and rhesus monkeys are presented and used, with the data of other investigators, in a comparison of Cercopithicini and Papionini species. An analysis of these data suggests that the Cercopithecini are characterized by a general behavioral strategy that differentiates this tribe from the Papionini just as clearly as do morphological and chromosomal characteristics. The elucidation of such distinctions does more, however, than add to an understanding of Cercopithecinae systematics. The observed behavioral variability carries potential implications also for the interpretation of dominance phenomena and for our understanding of the contribution of phylogeny to social structure. Hypotheses relating to these issues are offered in the hope of provoking rejoinders and further research by other investigators.

METHODS

Behavioral data were derived in large part from four studies, two on *Macaca mulatta* and one each on *E. patas* and *C. aethiops*. The behavioral study sites, numbers of monkeys observed, and hours of observation are shown in Table I. The patas monkeys shared an island with two small groups of rhesus monkeys, while the vervets at the Yerkes facility were housed for the majority of the study with a large group of sooty mangabeys (*Cercocebus atys*). In addition to these locations, brief observations were also made (on vervets) at the Behavioral Sciences Institute on St. Kitts, West Indies, and on the guenons, drills, and mandrills housed at the Birmingham (Alabama) Zoo.

TABLE I. Identification and Location of Study Groups

Species	Hours observed	Location	Adult males	Adult females	Immatures
C. aethiops[a]	58	Yerkes Regional Primate Center Field Facility (¼-acre enclosure)	4	2	3
E. patas[b]	400	La Parguera, Caribbean Primate Center (Guayacan, 94-acre island)	1	18	23
M. mulatta[c]	51	Yerkes Regional Primate Center Field Facility (¼-acre enclosure)	16	22	35
M. mulatta[d]	1530	Cayo Santiago, Caribbean Primate Center (40-acre island)	24	27	43

[a]Maintained together with a large group of Mangabeys. This group of vervets had been in place for several years. A single breeding pair gave rise to five of the seven other animals in the group.

[b]See Kaplan and Zucker [1980] for a history of this group.

[c]Supplemented by several hundred hours of observation by D. Chikazawa. This group of monkeys had been formed several years earlier; most of the adults were feral born.

[d]Additional information available in Kaplan [1977].

The same data collection techniques were utilized for the three species shown in Table I. For purposes of this report, only a few of the interaction patterns will be described. These behaviors include patterns of dyadic fighting (and the resulting dominance relationships and hierarchies) and triadic fighting (''fight interference''). In addition, aspects of affiliation (grooming, spacing) are described.

The direction and outcome, as well as frequency, of all observed fights were recorded. Following the definition of Sade [1967], fights were interactions in which an attack of any intensity was followed by flight of any intensity. Dominance was determined on the basis of fight outcomes only [Sade, 1967]. Animals eliciting flight gestures during fights were judged dominant to animals displaying such gestures. Where outcomes were uncertain, the last animal to have fled the fight was considered to be the probable winner. That is, the animals that ''displaced'' other animals were considered to be probable winners. This measure was especially useful in the Cercopithecini, where subordination through spatial avoidance seemed particularly frequent. These episodes were scored separately from clear fight outcomes.

Triadic fights refer to those incidents in which a third monkey entered a fight that began between two others (an "attacker" and a "victim"). In the data to be described here, all such instances involved aid to one of the combatants (the "beneficiary") at the expense of the other combatant (the "target") [Kaplan, 1977]. In a minority of instances, especially in the corral situations, a large number of animals would be involved in coalition or fight interference behavior, making triads impossible to discern. The data described here reflect only those instances that contained clearly defined triadic associations. Episodes with more than three combatants and triadic fights in which the beginning was not seen were not included.

Affiliative behavior is described in terms of grooming and passive body contact. Not all aspects of grooming are described, only the frequency and length of the grooming episodes. Passive body contact included those instances in which animals were calmly sitting in contact with no other behavior between them.

The analysis of some of the data was complicated by the fact that the free-ranging monkeys (rhesus and patas) were observed with the use of ad libitum samples only, while at the Yerkes Center, the monkeys were observed with both ad libitum and focal samples [Altmann, 1974]. Rates of performance of behavior are easily calculated from the focal samples, while rates determined from ad libitum samples are only indices of performance. These latter rates were determined as follows. First, the number of episodes of a particular type was divided by the number of hours over which the data were collected. Next, the average number of animals visible at any one time (determined through the use of scan samples) was divided into this number to yield an episode per hour per visible monkey. This was done both for the rhesus monkeys at Cayo Santiago and the patas monkeys at La Parguera. For comparative purposes only, an ad libitum "index" was created for the animals living at the Yerkes facility. This was done by dividing the number of episodes of a particular type by the number of hours of observation over which the data were collected. For the rhesus monkeys, this number was divided by the number of monkeys estimated to be within the observer's direct view at any one point in time (n = 12). The number 12 was derived from a series of random scan samples in which those individuals acting as the focus of the observer's attention at that time were identified and counted. For the purposes of this ad libitum estimate, it was assumed that all of the vervets (n = 9) were visible at all times.

All behavioral comparisons were made between species living under the same conditions. That is, the free-ranging patas were compared with free-ranging rhesus monkeys, while the vervets in the compound at Yerkes were compared to similarly housed rhesus monkeys. Most statistical comparisons utilized the proportion of acts performed by each species. These proportions were compared with the use of a "z" statistic [Bruning and Kintz, 1977]. The comparison involving length of grooming episodes utilized t-tests. All tests of significance were two-tailed.

In addition to behavioral data, comparative information on deaths associated with fighting are presented. These data were collected from observations and breeding records of guenons [de Brazza's monkeys (*Cercopithecus neglectus*), diana monkeys (*C. diana*), Sykes' monkeys (*C. mitis*), spotnosed monkeys (*C. nictitans*)], mandrills (*Mandrillus sphinx*), and drills (*M. leucophaeus*) in a zoo environment (Birmingham, Alabama), and from the veterinary records of the vervet monkey and cynomolgus macaque (*Macaca fascicularis*) breeding colonies maintained at a research facility (Bowman Gray School of Medicine).

RESULTS

In general, behavioral data from the four major study groups were consistent with those described by observers working with the species under more naturalistic conditions. Behavior in the corrals was, of course, truncated to a certain extent by physical restrictions placed on individuals and groups and by the absence of predators or other extra-group threats.

Dyadic Fight Patterns

Agonism was commonly observed in all the study groups (Table II). Direct comparison of rates of agonism may be made between the two free-ranging groups (based on ad libitum samples and corrected for observability) and between the two corral groups (based on focal samples). These data indicate that the African species in both locations had equivalent or lower rates of agonism than rhesus monkeys. Comparing the data based on the ad libitum samples, it seemed that the corral setting was accompanied by higher rates of agonism among the rhesus monkeys than in the free-ranging situation, while such a pattern of change was not apparent for the African species.

A striking difference between the species was observed in the use of cowers and grimaces as terminators of fights (Table II). Most of the fights among the rhesus monkeys (approximately 85%) had outcomes that could be determined clearly based on directed cowers and grimaces. Among the patas, and particularly among the vervets, fights tended to end with the animals merely moving apart or with one moving away and giving no clear gesture of submission. This difference between the African monkeys and between

TABLE II. Patterns and Rates of Fighting

Group	No. fights	No. bites	Rate[a] (ad lib)	Rate[b] (focal)	Fights having clear terminations[c]
M. mulatta (Cayo Santiago)	9684	370	0.63	—	8230
E. patas (La Parguera)	1424	87	0.56	—	704
M. mulatta (Yerkes)	753	76	1.23	5.2	648
C. aethiops (Yerkes)	249	3	0.48	2.2	87

[a]Per hour per monkey, estimated for monkeys at the Yerkes facility as described in methods.
[b]Per hour per monkey.
[c]As indicated by cower or grimace.

the macaques was significant (rhesus vs patas: $z = 22.4$, $p < 0.001$; rhesus vs vervets; $z = 15.7$, $p. < 0.001$).

Another difference in the pattern of fighting among the species involved the signalling of social subordination in the absence of an overt attack. Among the rhesus monkeys, 25% of all fights involved clear gestures of social subordination without a preceding overt attack. One animal would merely move near or move toward another, and this was enough to elicit a cower or grimace. Among both the patas and vervets, only about 10% of all fights involved such apparently unprovoked subordination (rhesus vs vervets: $z = 5.0$ $p < 0.001$; rhesus vs patas: $z = 93$, $p < 0.001$).

Finally, fighting among the rhesus monkeys was marked by a relative absence of double attacks (i.e., both participants in a dyadic fight displaying attack gestures). Among the rhesus monkeys this occurred in less than 3% of all fights, while among the patas it occurred in about 10% of all fights ($z = 5.89$, $p. < 0.001$). Double attacks were more common numerically among the vervets than among the rhesus, but the difference was not significant.

Despite the differences in the relative frequency of clear terminations of fights, the overall pattern of dominance relationships within groups was similar across species. In each instance, a linear hierarchy was present among adult females, with few entries below the diagonal (less than 1% in all cases). Such hierarchies were also observed among rhesus and vervet males. Thus, although there were more uncertain and unclear fight outcomes among both the patas and the vervet monkeys, most outcomes were consistent with a linear hierarchy, similar to the pattern observed for rhesus monkeys. Interestingly, while the dominant animal in the vervet group was a male, the second ranking animal was an adult female. She was ranked above her mother and three older adult males, two of which were half brothers. The dominant animal in both rhesus groups and in the patas group were adult males. The second ranking animals in the rhesus groups were also males. There was only one adult male in the patas group; the second ranking animal was thus a female. A second adult male was observed occasionally in the vicinity of the patas group, but his presence during observations usually evoked an agonistic response by group members.

Triadic Fights

Triadic fights (''fight interference'') were episodes in which a monkey entered a fight that began between two others. This behavior also has been characterized as ''policing'' [Rowell, 1971] or ''alliancing'' [Massey, 1977], but the term ''interference'' probably has precedence as it was first used by Schjelderup-Ebbe in his descriptions of fights among barnyard fowl in 1922 [1935, 1977]. In virtually all instances in which the activities of all participants could be observed, aid was given to one of the combatants at the expense of another.

The pattern of fight interference or fight aiding is shown in Table III. Rhesus monkeys interfered more actively in fights than either of the African species (rhesus vs patas: $z = 8.60$, $p < 0.001$; rhesus vs vervets: $z = 6.50$, $p < 0.001$). Perhaps more important than the relative frequency of the interference or fight aiding behavior was the marked tendency for the interferer among rhesus monkeys to aid the victims of aggression while interference by the African species was characterized by aid to the aggressor, not the victims (rhesus vs patas: $z = 6.8$, $p < 0.001$; rhesus vs vervets: $z = 3.38$, $p < 0.001$).

Aid to relatives versus nonrelatives is also shown in Table III. The information is not presented for the patas monkeys, since their genealogical relationships were unknown except for mother-infant pairs. However, it is suspected that interference in patas monkeys was more like that of vervets (i.e., low frequency of aid to relatives) than it was like that of the rhesus monkeys (high frequency of aid to relatives; rhesus vs vervets: $z = 3.4$, $p < 0.001$). This is because most of the patas interference aid (75%) went from adult females to other adult females, and many of these animals were probably unrelated [Kaplan and Zucker, 1980]. Relatively little aid went from adults to infants and juveniles, the animals most likely to be related to the adults. In contrast, Table III shows that among the rhesus at Cayo Santiago and at the Yerkes facility, a much higher percentage of aid went to juveniles and infants (53% and 43%, respectively). Finally, the adult male

TABLE 3. Fight Interference Behavior

Group	No. interferences	% total fights	Aid victim	Aid attacker	Aid relative
M. mulatta					
(Cayo Santiago)	987	10	721	266	634
E. patas					
(La Parguera)	41	3	14	27	—[a]
M. mulatta					
(Yerkes Field Station)	186	25	104	82	125
C. aethiops					
(Yerkes)	16	6	2	14	4

[a]Unlikely to have been more than 25%, as 75% of aids were between unrelated adults.

rhesus monkeys were much more active in aiding than either the patas or vervet males. Over 30% of all aids among the rhesus monkeys were performed by adult males versus 3% for the patas and vervets combined ($z = 4.38$; $p < 0.001$).

Male-Male Relationships

An important aspect of nonhuman primate social behavior concerns the relationship between males. Among vervets, male-male relationships tended to be antagonistic, as indicated by the frequency with which males would attack each other. One pattern of attack and harassment involved what Struhsaker [1967a] has called the red, white, and blue display, in which a dominant animal prances back and forth in front of a subordinate while displaying his brightly colored perianal region. Among animals living at the Yerkes facility, displays involving a perianal presentation were not observed. Rather, observations revealed a display involving a frontal approach by a dominant animal followed by head weaving and bobbing in front of a subordinate target ($n = 51$ instances). The target monkey often responded with a cower or a crouch and a facial expression in which the mouth was open, the lips were retracted and a "eh eh" or cackle vocalization was emitted. The display appeared to be controlled by the dominant animal and accompanied by nervousness and agitation by the target animal. Further, this display seemed to involve a high degree of harassment of the target by the dominant animal.

Occasionally, when the harassing animal was the dominant male of the group, an interesting variation of this behavior occurred. The "harassed" animal would begin following the dominant animal, which then began loping away. Other animals might join in this "chase" until virtually the entire group was involved. The animal so chased invariably set the rhythm and speed of the chase. In the 21 chase instances observed among the vervets, the dominant animal was never caught; in fact he himself ended the "chase" by sitting down. Nothing comparable to this male "harassment" display was observed among the rhesus monkeys. In the patas monkey group there was only one adult male. However, a display given by young males involving a crouch and batting response to the approach of the dominant male was not uncommon (n = 50 instances) [Zucker and Kaplan, 1981]. The male typically ignored the crouching and batting of the juveniles. The motor patterns displayed by these young patas males were reminiscent of behaviors shown by harassed vervet males. The function of such displays remains obscure, although at least in the case of the vervets, it may have had a cohesive or signalling function since, at times, the entire group would become involved.

A second aspect of male-male interaction concerns copulation interference. Among the rhesus monkeys, the dominant male or males sometimes tried to interfere with matings occurring within their field of vision. In my observations (of over 100 instances), however, the objects of these attacks were almost always (> 95%) the female member of the mating pair. At Yerkes, fewer than five matings were observed among the vervets. In the one copulation in which interference by the dominant male was observed, it was the male of the copulating pair that was attacked. Matings among mangabeys or attempts by vervet males to mate with mangabey females were more commonly observed (e.g., 12 instances of vervet-mangabey matings). On several occasions, the dominant male vervet attacked male members of mangabey consorting pairs, and one time attacked another adult male vervet who was attempting to mate with a mangabey female. Brief observations on vervets in a corral on St. Kitts during the mating season of 1976 contained six instances of males breaking up consorting pairs; in each case it was the male of the mating pair that was attacked. It was the willingness of the vervet males to engage in direct competition and aggression with other adult males that appeared to mark them as distinct from adult male rhesus monkeys, which appeared to avoid such interactions.

Observations on adult patas male-male interactions were rare. A nongroup adult male was frequently observed alone at La Parguera. As reported above, his presence in the vicinity of the group was greeted by alarm, agitation, and aggression by the harem male [Kaplan and Zucker, 1980].

Grooming Behavior and Passive Contact

The frequency and length of grooming episodes were compared among the three species, and these showed significant differences (Table IV).

TABLE IV. Patterns and Rates of Grooming and Body Contact

Group	Grooming episodes	Rate: Ad lib samples[c]	Rate: Focal samples[d]	Length[a] in minutes (± SEM)	Body contact episodes[b]	Rate: Ad lib samples[c]	Rate: Focal samples[d]
M. mulatta (Cayo Santiago)	30,109	1.52	—	3.66 ± 0.42	7,740	.389	—
E. patas (La Parguera)	446	0.177	—	1.87 ± 0.11	5	.002	—
M. mulatta (Yerkes)	861	1.41	5.17	2.58 ± 0.43	199	.325	5
C. aethiops (Yerkes)	230	0.44	0.91	0.88 ± 0.08	86	.165	0.4

[a]Timed episodes: n = 71 (Cayo Santiago); n = 292 (La Parguera); n = 50 (Yerkes rhesus); n = 162 (Yerkes vervets).

[b]Affiliative contact in which participants are passively touching, with no other behavior (such as grooming) occurring.

[c]Per hour per monkey. See Methods for method of estimating rates from scan samples at Yerkes.

[d]Per hour per monkey.

Estimates of grooming rates based on ad libitum samples indicate that grooming was a frequent activity among rhesus monkeys but relatively infrequent among vervets and especially among patas monkeys. The focal samples indicate rhesus monkeys in compounds groomed about six times as frequently as vervets. Under free-ranging conditions, rhesus monkeys groomed about eight times as frequently as comparably observed patas monkeys. In addition to differences in grooming frequency, the length of grooming episodes differed between groups. The grooming episodes of rhesus monkeys were significantly longer than those of comparably housed vervets (t = 3.89, $p < 0.01$) and patas monkeys (t = 4.12, $p < 0.01$).

At both the Caribbean Primate Center and at the Yerkes Regional Primate Research Center, the African monkeys engaged in many fewer episodes involving passive (affiliative) physical contact (Table IV). The patas monkeys almost never were observed to engage in this behavior. Upon superficial examination, vervets would sometimes appear to be in physical contact; closer examination with binoculars would often reveal the animals to be maintaining several cm distance.

TABLE V. Deaths Due to Trauma in Adult Females Living in One-Male, Multifemale Groups at the Bowman Gray School of Medicine

Species	Years	Female-years[a]	Deaths due to trauma[b]
C. aethiops	1978–1982	151	14
M. fascicularis	1978–1982	457	13

[a]To derive the number of female-years, the total number of days each female was present in a one-male harem was summed across females and divided by 365.

[b]$z = 3.33$, $p < 0.001$, comparison based on the ratio of trauma deaths to female-years.

Deaths Associated With Fighting (Trauma)

The data on death due to fighting derived from cynomolgus macaques and vervet monkeys maintained in one-male, multifemale breeding groups at the Bowman Gray School of Medicine are shown in Table V. It can be seen that despite the larger number of females at risk, the cynomolgus macaques suffered fewer deaths that could be attributed to trauma than did the vervets, when both were housed under similar conditions ($z = 3.33$; $p < 0.001$).

A second set of comparative data was derived from observations made on guenons, drills, and mandrills living in small ($n = 3$ to 8 animals) family groupings in indoor-outdoor pens at the Birmingham Zoo. These observations must be interpreted cautiously, as a relatively small number of animals was involved. There were four species of guenons at the zoo, each in a single family grouping consisting of an adult male and female, and one or more offspring: 1) diana monkeys (*C. diana*); 2) de Brazza's monkeys (*C. neglectus*); 3) Sykes' monkeys (*C. mitis*) and 4) spotnosed monkeys (*C. nictitans*). Over a 6-year period three diana monkeys and one de Brazza's monkey were killed by family members. Additionally, members of the Sykes' group had to be separated after serious fighting occurred between siblings. In addition to these killings, relatively frequent wounding among all the species was reported in the colony records.

Apparently, neither long tenure together nor genealogical affiliation limited serious aggression within these confined, family groups of guenons. The guenon observations contrast with those made on a family group of drills ($n = 8$) and a family group of mandrills ($n = 5$) (*Mandrillus leucophaeus* and *Mandrillus sphinx*), living in equivalently-sized pens at the same zoo. There was no record of conspecific killing or serious wounding in these groups despite the fact that the drills and mandrills were more crowded (because of their larger size) than the guenon groups.

Comparative Data

The data reviewed above suggests a set of consistent differences between the Cercopithecini and the Papionini in the way each group expresses and regulates interindividual agonistic and affiliative contacts. These results, especially those relating to the vervets, must be interpreted conservatively, as the social context was limited. However, data reported by other investigators suggest a similar set of differences between species of the two tribes.

Dyadic Agonism

A pattern of dyadic fighting peculiar to the Cercopithecini (including talapoin monkeys) is reported by several investigators. Three major elements of this pattern are: 1) a marked lack of unidirectionality and asymmetry in fights [Bernstein, 1970; Rowell, 1974; Gartlan and Brain, 1968]; 2) the lack of ritualized gestures of submission and appeasement including the fear grimace (delivered at a distance in the absence of overt aggression) and the social present [Wolfheim and Rowell, 1972; Bolwig, 1978; Gartlan and Brain, 1968; Struhsaker, 1967a; Oswald and Lockard, 1980]; and, 3) a tendency for persistent attacks, growing progressively more severe with time, with serious wounding or even death not uncommon [Bramblett et al, 1982; Bolwig, 1978; Gartlan and Brain, 1968; Rowell, 1971; Wolfheim and Rowell, 1972]. These patterns of agonistic behavior are apparently most obvious under conditions of captivity (corrals, small groups cages) as a low level of aggression and a high degree of avoidance are reported for these species in the wild [Hall 1965; Rowell, 1974]. On the other hand, most of the Papionini (viz, macaques and baboons) are reported to be highly aggressive both in the wild and in captivity [Chalmers, 1980]. Their aggressiveness, however, tends to be asymmetrical and unidirectional, highly ritualized, and not often related to progressive wounding or killing [Hausfater, 1975; Rowell, 1966, 1971]. This is not to suggest that such outbursts never occur in macaques and baboons [*cf.* Samuels and Henrickson, 1983; Goo and Sassenrath, 1980], only that they are, perhaps, less frequent.

Triadic Agonism

Polyadic interactions comprise another aspect of agonism. Interference in fights ("policing behavior," "agonistic aids," "control") has been described in detail in a series of studies on Papionini species [Kummer, 1967; Tokuda and Jensen, 1968; Oswald and Erwin, 1976; Massey, 1977; Kurland, 1977; Kaplan, 1977, 1978; Cheney, 1977; Berman, 1980; Bernstein and Ehardt, 1985]. Many functions have been ascribed to this behavior, including the role it may play in the acquisition and maintenance of dominance ranks within the group. Also noted frequently is the importance of genealogical relationships in predicting its occurrence. Finally, an aspect of fight interference noted by many observers working with macaques and baboons is the tendency for interference to be of benefit to fight victims rather than aggressors.

In general, interference behavior has not been as thoroughly described for the Cercopithecini. Thus, Struhsaker [1967b] reported coalitions in about one-fifth of the fights among the vervets, but it is not clear if such coalitions were offensive (directed against original victims of aggression, as reported above for vervets and patas) or defensive (directed against aggressors). Seyfarth [1980] described coalitions among vervets as largely involving top ranking females helping each other. These episodes may well have been offensive, as lower ranking members of groups were probably the ones being attacked as a result. Fairbanks [1980] noted the relative absence of coalitions among adult vervets, while Rowell [1971] and Bolwig [1978] specifically referred to the absence of defensive alliances or policing behavior among guenons, emphasizing instead the tendency of animals to ''gang up'' on victims. More recently, Horrocks and Hunte [1983a] reported on interference behavior of vervet mothers on behalf of their offspring. They found mothers most often intervened when dominant to the target animals and when their own offspring were the aggressors. Their data also suggested such interventions to be relatively infrequent in comparison to the number of fights. In sum, the element that appears most often in reports of interference behavior among the Cercopithecini is the tendency for aggressors rather than victims of aggression to be aided.

Male Role

Other reported differences between the Cercopithecini and the Papionini concern the social role of males and the quality of male-male interaction. In particular, there are numerous reports in the literature of Cercopithecini adult males being socially peripheral to their groups and highly antagonistic to each other, in contrast to most reports of behavior among the Papionini males. The tendency for adult males to be peripheral and not involved with the social activities of the group (including the alarm barks and the control of aggression) has been reported for a variety of *Cercopithecus* species, as well as for *Erythrocebus* [Struhsaker, 1969; Gartlan, 1974; Dunbar, 1974; Byrne et al, 1983]. Among vervets and talapoins, in particular, relationships between adult males and adult females are described as negative, occasionally involving high levels of aggression [Krige and Lucas, 1975; Rowell, 1973]. The more peripheral social role of the Cercopithecini males is further emphasized by the frequent observation that adult female vervets are often dominant to adult males and are sometimes the most dominant members of the group [Bramblett et al, 1982; Bernstein, 1970; Lancaster, 1971]. In talapoin groups, adult females have been known to kill adult males [Wolfheim, 1977]. The presence of a female as the most dominant animal is not reported nearly as frequently for groups of macaques and baboons.

Intense Cercopithecini male-male antagonism and conflict has also been reported by several investigators, even among vervets, in which multimale groupings are often observed [Struhsaker, 1977; Henzi and Lucas, 1980; Bourliere et al, 1970; Dunbar, 1974]. There are no reports of male coalitions and central hierarchy formations for vervets as have been reported for macaques and baboons [Kaufmann, 1967; Hall and Devore, 1965]. The willingness of males to engage in other forms of direct competition (e.g., in copulation harassment) has not often been documented. However, Henzi and Lucas [1980] indicated that male vervets attack the male members of copulating pairs. In contrast, Hausfater [1975] suggested that the female baboon is typically the recipient of male copulation harassment, as suggested by the above data for rhesus monkeys.

Grooming and Spacing

Patterns of affiliation are described by many investigators as being markedly different in the Cercopithecini than in the Papionini. Studies of patas monkeys [Struhsaker and Gartlan, 1970], vervets [Bernstein, 1970; Fairbanks et al, 1978], and other guenons [Struhsaker, 1969; Bolwig, 1978; Haddow, 1952; Oswald and Lockard, 1980] suggest that grooming among adults is *relatively* infrequent. These comments do not mean that the Cercopithecini do not groom; they do, and sometimes intensively [e.g., Hall's (1965), description of patas grooming during the birth season, and Seyfarth's (1980) description of grooming among adult female vervets]. However, comparative observations often indicate grooming to be more frequent and perhaps more evenly distributed among group members in the Papionini than Cercopithecini species. For example, in Seyfarth's study of vervet females [1980], over 40% of all grooming was both initiated and received in the 20% of dyads representing the high ranking animals of his three study groups.

Proximity (interindividual distance) is another index of affiliation. Several investigators suggest that distance between individuals is greater among individuals of Cercopithecini species than among the Papionini, observations that are consistent with the lower frequencies of grooming sometimes reported [Struhsaker and Leland, 1979; Dunbar, 1974; Kummer, 1974;

Bolwig, 1978; Oswald and Lockard, 1980; Haddow, 1952; Rowell, 1971]. No doubt there is variability among the Cercopithecini species in the degree of interindividual spacing. For example, Schapiro (personal communication) reports that adults in captive one-male multifemale groupings of blue monkeys, Sykes' monkeys, and vervets spend relatively little time in contact. However, among these animals, vervet adults were significantly more likely to be in contact than adults of these other two species.

Genealogical Associations

Patterns of affiliation (grooming, spacing, aid in fights) among some of the Papionini are reported to be influenced strongly by genealogical relationships [Kawai, 1958; Sade, 1965, 1967; Massey, 1977; Kurland, 1977; Kaplan, 1977, 1978; Berman, 1980; Bernstein and Ehardt, 1985]. Among Japanese and rhesus macaques and baboons there has been described a well-defined genealogy-dependent algorithm predicting the integration of young females into the adult female dominance hierarchy [Kawai, 1958; Sade, 1967, 1972; Chikazawa et al, 1979; Hausfater et al, 1982].

The data for proximity, grooming, and fight aiding or interference reported above for the vervets and patas monkeys, and the patterns of intraspecific killing reported among the captive groups of guenons, do not indicate a particularly strong effect of genealogical relationships on patterns of affiliative interactions, except perhaps, within the mother-infant dyad. Two features may be relevant in gauging the strength of genealogical associations within Cercopithecini groups: 1) the pattern with which juveniles attain independence from their mothers; and 2) the manner in which juveniles, especially females, become integrated in the adult dominance hierarchy.

Regarding juvenile independence, many reports indicate that patas monkeys and vervets experience independence from their mothers relatively early in life [Krige and Lucas, 1975; Lancaster, 1971; Rowell and Richards, 1979; Struhsaker, 1971], especially compared to macaques and baboons [Seay et al, 1970; Baker and Preston, 1973; Fairbanks, unpublished]. It might be speculated that such early independence presages relatively weak genealogical bonding later in life. In this context it is worth noting that juvenile baboons (up to 36 months) more actively engage in both affiliative and agonistic interactions than do similarly-aged Sykes' monkeys and vervets [Bramblett and Coelho, this volume].

Several studies have reported on the relationship between maternal dominance rank and rank of offspring among the Cercopithecini. For example, Horrocks and Hunte [1983b] and Loy and Loy [this volume] showed that, as among macaques, the ranks of juveniles in dominance hierarchies of peers follow closely the relative ranks of their mothers. However, there has been only one study in which the ranks of juveniles were followed longitudinally [Bramblett et al, 1982]. In this study it was clear that the ultimate (adult) rank of daughters was influenced significantly by the rank of their mothers. However, the pattern of sister reversal and the rise in rank of youngest daughters to a level just below that of their mothers was not as clearly expressed as in the macaque groups described by Kawai [1958] and Sade [1967]. In fact, the correlation between the dominance ranks of vervet daughters and their mothers, while significant (r_s = .40) was low in comparison to that observed in published data for macaques [r_s = 0.90; Sade, 1972]. Further, while Horrocks and Hunte [1983b] described sister reversal among juvenile vervets, their data were not longitudinal and thus did not show what the ranks of these juveniles were as they reached adulthood. Horrocks and Hunte [1983a] also suggested that rank acquisition of juvenile vervets depends strongly on the pattern of aggression initiated against juveniles by females other than their mothers. In contrast, these authors suggested that among macaques "passive" acceptance of rank and maternal intervention play a relatively larger role in rank acquisition.

While only suggestive, the data on infant development and rank acquisition do indicate that the importance of, or at least the patterning of, genealogical relationships may differ significantly between the Cercopithecini and Papionini. Clearly, these issues only will be settled with more longitudinal data from Cercopithecini species, especially from species other than vervets.

DISCUSSION

The data presented and the literature reviewed suggest that members of the Papionini and Cercopithecini tribes differ in the patterning or expression of social behavior. These differences may be important for what they imply about basic concepts in primate biology, such as "dominance," "hierarchy," and the relative contribution of phylogeny to social organization [Gartlan, 1973]. These points are discussed below.

Dominance

Wade [1978] described the salient components of a well-developed system of agonistic status (dominance) determination: 1) a large repertoire of appeasement and submission gestures, often used in ritualized fashion to establish or signal status relationships; 2) a marked ability to learn (and accept) status differentiations; and 3) a well-developed social acumen, allowing animals to use the behavior of others to alter their own status or to insulate themselves from the effects of the status of others.

Of these features, the essence of dominance is perhaps best described in terms of 2) —the ability (or willingness) to learn hierarchical (asymmetric) relationships. Not only must the loser of the fight be willing to submit, but the winner must accept the submission and not escalate the fight.

The data reviewed above suggest that the Cercopithecini may not have a particularly elaborate system of status determination. Thus, ritualization of fighting in the Papionini via an elaborate repertoire of appeasement and submissive motor patterns and facial expressions is largely absent in the Cercopithecini. Further, the willingness to form complex coalitions for defense apparently is also minimal or absent among the Cercopithecini.

The *relatively* nonritualized nature of dominance relationships among the Cercopithecini, as compared to the Papionini, suggests that ''dominance'' (i.e., the establishment and maintenance of asymmetrical hierarchical relationships) has variable importance and function across Old World monkey taxa. It appears to be subject to the same taxonomic distinctions as ischial collosities, neck bites, cheek pouches, and canine lengths. That is, dominance patterns, like these other aspects of morphology and behavior, enter into the adaptive equation by which monkeys confront their environments. The adaptive equation of the Cercopithecini relative to that of the Papionini apparently does not involve an elaborate, ritualized social dominance system. Under natural conditions, members of most Cercopithecini species are not particularly aggressive, and depend on spatial resolutions to most conflicts [Rowell, 1971]. That is, changes in spatial relationships have been substituted for submissive facial expressions and postures, and for defensive coalitions. This would imply a social structure that uses space, rather than one that conserves space.

In contrast, the relatively well-developed status networks and patterns of interaction among the Papionini may be related to the ability these networks confer on animals to establish spatially compact, and behaviorally coordinated groupings. These are groupings in which animals can survive despite frequent and repeated conflict over space, food, mates, and the other desiderata of monkey life. Perhaps these species are, as a result, charged with an aggressive potential that makes possible both effective predator defense and displacement of competing species from a broad range of environments [Altmann and Altmann, 1970]. In this view, the advantage of a well-developed network of social dominance is not necessarily that it reduces aggression but that it controls the severity of aggression with ritualized patterns of submission, making life within a highly aggressive society possible. Individual success (e.g., survival) in such a society depends, at least in part, on the willingness and ability of animals to recognize dominance relationships and to order their social activities (agression, submission, alliance formation) around such relationships. The outcast (whether of high or low social status) will be the animal who cannot recognize or accept its status.

It would be predicted from these hypotheses about Cercopithecini and Papionini social systems that Cercopithecini species would adapt less successfully to environments lacking adequate cover or space or to environments involving social disruption through introduction and removal of animals. The evidence from the literature is consistent with these suggestions, as captive group formation and maintenance tend to be relatively more problematic in the Cercopithecini than in the Papionini [Rowell, 1971; Bolwig, 1978; Kaplan et al, 1985]. Again, this does not imply that among macaques and baboons group formation or the introduction of strangers occurs without serious aggression [Bernstein et al, 1974; Goo and Sassenrath, 1980]. However, it is argued that such aggression is less severe and generally more rapidly resolved than would be observed among similarly treated Cercopithecini groups (e.g., Table V, which contains data from periods of group formation among vervets and cynomolgus monkeys).

Hierarchy

Many investigators distinguish between dominance relationships (the formation or existence of an asymmetrical, hierarchical agonistic relationship between two individuals) and dominance hierarchies (the ordering of animals in the group or subgroup on the basis of relative dominance) [Wade, 1978; Bernstein, 1981]. In the current review, the emphasis has been on relationships, and not on the resulting hierarchies or on the ranks animals may occupy within those hierarchies. Linear and stable dominance hierarchies are found in both tribes, apparently to an equivalent extent [Bramblett et al, 1982]. The relatively unvarying nature of the hierarchies suggests that they are not units of selection, but rather are a result or outcome of interindividual competition [Count, 1958]. It is the manner and ease with which asymmetric, hierarchical dominance relationships are established, and their subsequent role in ordering other aspects of social life, which appears to be a critical element and the object of selection. As suggested above, the Papionini and the Cercopithecini may differ in the production of these relationships.

These statements should not be construed as suggesting that dominance hierarchies are irrelevant in the socioecology of monkeys, or that the positions of animals within these hierarchies are unimportant. Indeed, several recent studies indicate that in both tribes, rank positions in a dominance hierarchy may predict the partitioning of resources and/or reproductive success within groups [Wrangham, 1981; Cheney et al, 1981; Wilson et al, 1978; Sade, 1980; Sapolsky, 1982; Adams et al, 1985; see Strum, 1982, for a contrary view].

Conclusions: Phylogeny and its Potential Contribution to Social Structure Within the Cercopithecinae

Count [1958] suggested that the lifeway or biogram of a taxon was composed of two elements; 1) a "heritage" defined by those characteristics shared with related taxa at a higher level (e.g., a mammalian heritage held in common by all families of primates); and 2) a "habitus," defined by those characteristics distinguishing a particular taxon from related taxa at the same level (e.g., a Cercopithecine habitus vs a Colobine habitus). In Count's formulation of the biogram, behavioral components are assumed to be as relevant taxonomically as anatomical and physiological components. This is because all of these elements presumably arise from the same evolutionary and ontogenetic processes.

The data presented and reviewed in this essay suggest a habitus for the Cercopithecini that distinguishes them from the Papionini and that involves three major aspects: 1) a relatively limited social role for males; 2) a reduced or underdeveloped ability to establish and maintain ritualized hierarchical relationships; and 3) infants that are relatively precocial at birth. These factors correlate with a tendency for the development of relatively small one-male groups in which perhaps social relationships are modulated by spatial adjustments and in which genealogical ties remain relatively undeveloped. These are all characteristics that tend to be represented in opposite fashion among the Papionini.

According to this view, the talapoins [considered by Jolly (1966) to be a tribe *incertae cedis*] are in fact members in good standing of the Cercopithecini. Furthermore, *Erythrocebus patas* should not be considered an aberrant species in the tribe Cercopithecini but as a representative of the logical extension of Cercopithecini characteristics to a terrestrial niche. This view also suggests the importance of phylogenetic considerations in understanding the distribution and expression of social behavior across taxa. Thus, for example, vervets, macaques, and some species of baboons are often categorized together as dwelling in "multimale" groupings or "female bonded" groupings [Wrangham, 1980]. The information in this essay implies that such structural similarities may represent different underlying behavioral processes.

Further insight into the specific attributes embodied in such behavioral processes awaits detailed studies from a broader range of species, especially among the Cercopithecini. To this point, the vervets have been the most thoroughly studied species of this tribe and thus provide the basis for the most of the generalizations made about the Cercopithecini. In addition to behavioral studies, useful information regarding similar and contrasting modes of existence might well be provided by studies examining species-typical physiological responses to behavioral challenges. Recent work on New World monkeys by Mason, Mendoza, and their colleagues [e.g., Mendoza and Moberg, 1985; Mason, 1985] suggests that interspecies differences in social behavior and "temperament" are accompanied by differences in physiological responses to psychological and social challenges. These investigators indicate that such interspecies differences in physiological responsiveness may be used as additional evidence concerning phylogenetic relationships. If some of the hypotheses presented above are correct, behavioral contrasts observed between Papionini and Cercopithecini species could well be associated with differences in adrenocortical, serotonergic, immunologic, or cardiovascular responses to psychogenic challenges [Kaplan, 1986].

ACKNOWLEDGMENTS

The data on which this chapter is based were collected, in part, with the support of grants from the NSF (GS 2377), NIMH (Predoctoral Grant 1 FO1 MH50103, Cuan), and the University of Alabama in Birmingham. Preparation of the manuscript was supported in part by grants from the National Heart, Lung and Blood Institute (HL 14164, R01 HL 26561). The data contributions of Mary Anthony (Bowman Gray School of Medicine), Dennis Chikazawa (Yerkes Regional Primate Research Center), Randall Reid and Bob Truitt (The Birmingham Zoo), and Evan Zucker (Loyola University of New Orleans) are gratefully acknowledged. Additionally, the comments and unpublished observations of Steve Schapiro are gratefully acknowledged, as are the comments of Carol Shively, Stephen Manuck, and Katherine Kaplan; however, these individuals do not necessarily agree with the conclusions and interpretations presented in this chapter.

REFERENCES

Adams MR, Kaplan, JR, Koritnik DR (1985): Psychosocial influences on ovarian endocrine and ovulatory function in *Macaca fascicularis*. Physiol Behavior 35:935–940.

Altmann J (1974): Observational study of behavior: Sampling methods. Behaviour 49:227–267.

Altmann SA, Altmann J (1970): "Baboon Ecology," Chicago: University of Chicago Press.

Baker RP, Preston DG (1973): The effects of interspecies infant interaction upon social behavior of *Macaca irus* and *Erythrocebus patas*. Primates 14:383–392.

Berman CM (1980): Early agonistic experience and rank acquisition among free ranging infant rhesus monkeys. Int J Primatol 1:153–170.

Bernstein IS (1970): Primate status hierarchies. In Rosenblum LA (ed): "Primate Behavior Vol. I." New York, Academic Press, pp 71–109.

Bernstein IS (1981): Dominance: The baby and the bathwater. Brain Behav Sci 4:419–457.

Bernstein IS, Gordon TP, Rose RM (1974): Aggression and social controls in rhesus monkey (*Macaca malatta*) groups revealed in group formation studies. Folia Primatol 21:81–107.

Bernstein IS, Ehardt CL (1985) Agonistic aiding: Kinship, age, and sex influences. Am J Primatol 8:37–52.

Bolwig N (1978): Communicative signals and social behavior of some African monkeys: A comparative study. Primates 19:61–99.

Bourliere F, Hunkeler C, Bertrand N (1970): Ecology and behavior of Lowe's guenon (*Cercopithecus campbelli lowei*) in the Ivory Coast. In Napier JR, Napier PH (eds): "The Old World Monkeys." New York: Academic Press, pp 297–350.

Bramblett CA, Bramblett SS, Bishop DA, Coelho AM Jr. (1982): Longitudinal stability in adult status hierarchies among vervet monkeys (*Cercopithecus aethiops*). Am J Primatol 2:43–51.

Bruning JL, Kintz BL (1977): "Computational Handbook of Statistics, 2nd edition." Glenview, Illinois: Scott, Foresman, and Company.

Byrne RW, Conning AM, Young J (1983): Social relationships in a captive group of diana monkeys (*Cercopithecus diana*). Primates 24:360–370.

Chalmers N (1980): "Social Behavior in Primates." Baltimore: University Park Press.

Cheney DL (1977): The acquisition of rank and the development of reciprocal alliances among free-ranging immature baboons. Behav Ecol Sociobiol 2:303–318.

Cheney DL, Lee PC, Seyfarth RM (1981): Behavioral correlates of non-random mortality among free-ranging female vervet monkeys. Behav Ecol Sociobiol 9:153–161.

Chikazawa D, Gordon TP, Bean CA, Bernstein IS (1979): Mother-daughter dominance reversals in rhesus monkeys (*Macaca mulatta*). Primates 20:301–305.

Count EW (1958): The biological basis of human sociality. Am Anthropol 60:1049–1085.

Delson E, Andrews P (1975): Evolution and interrelationships of the catarrhine primates. In Luckett WP, Szalay FS (eds): "The Phylogeny of the Primates." New York: Plenum Press, pp 405–446.

Dunbar RIM (1974): Observations on the ecology and social organization of the green monkey, *Cercopithecus sabaeus*, in Senegal. Primates 15:341–350.

Dutrillaux B, Couturier J, Muleris M, Lombard M, Chauvier G (1982): Chomosomal Phylogeny of forty-two species or subspecies of cercopithezoids (primates, catarrhini). Am Genét. Vol. 2, pp 96–109.

Fairbanks LA (1980): Relationships among adult females in captive vervet monkeys: Testing a model of rank related attractiveness. Anim Behav 28:853–859.

Fairbanks LA (unpublished): Mother-infant contact and the development of independence in vervet monkeys (*Cercopithecus aethiops sabaeus*).

Fairbanks LA, McGuire MT, Page N (1978): Social roles in captive vervet monkeys (*Cercopithecus aethiops*). Behav Processes 3:335–352.

Gartlan JS (1973): Influences of phylogeny and ecology on variations in the group organization of primates. In Menzel EW Jr (ed): "Symposia of the Fourth International Congress of Primatology: Precultural Primate Behavior." Basel: Karger, Vol. 1, pp 88–101.

Gartlan JS (1974): Adaptive aspects of social structure in *Erythrocebus patas*. In Kondo S, Kankin M, Eharca A, Karvamura S (eds): "Symposia of the Fifth International Primatological Society." Tokyo: Japan Science Press, pp 161–171.

Gartlan JS, Brain CK (1968): Ecology and social variability in *Cercopithecus aethiops* and *Cercopithecus mitis*. In Jay P (ed): "Primates, Studies in Adaptation and Variability." New York: Holt Rinehart and Winston, pp 253–292.

Giusto JP, Margulis L (1981): Karyotype fission theory and the evolution of Old World monkeys. Bio System. Vol. 13, pp 267–302.

Goo GP, Sassenrath EN (1980): Persistent adrenocortical activation in female rhesus monkeys after new breeding group formation. J Med Primatol 9:324–334.

Haddow AJ (1952): Studies on an African monkey, *Cercopithecus ascanius schmidti matschie*. Proc Zool Soc Lond 122:297–394.

Hall KRL (1965): Behavior and ecology of the wild patas monkey, *Erythrocebus patas* in Uganda. J Zool 148:15–87.

Hall KRL, DeVore I (1965): Baboon social behavior. In DeVore I (ed): "Primate Behavior: Field Studies of Monkeys and Apes." New York: Holt, Rinehart and Winston, pp 53–110.

Hausfater G (1975): Dominance and reproduction in baboons (*Papio cynocephalus*). Contrib Primatol 7.

Hausfater G, Altmann J, Altmann S (1982): Long term consistency of dominance relations among female baboons. Science 217:752–755.

Henzi SP, Lucas JW (1980): Observations on the intertroop movement of adult vervet monkeys (*Cercopithecus aethiops*). Folia Primatol 33:220–235.

Horrocks J, Hunte W (1983a): Maternal rank and offspring rank in vervet monkeys: An appraisal of the mechanisms of rank acquisition. Anim Behav 31:772–782.

Horrocks JA, Hunte W (1983b): Rank relations in vervet sisters: A critique of the role of reproductive value. Am Naturalist 122:417–421.

Jolly CJ (1966): Introduction to the Cercopithecoidea with notes on their use as laboratory animals. Symp Zool Soc Lond 17:427–457.

Kaplan JR (1977): Patterns of fight interference in free ranging rhesus monkeys. Am J Phys Anthropol 47:279–288.

Kaplan JR (1978): Fight interference and altruism in rhesus monkeys. Am J Phys Anthropol 49:241–250.

Kaplan JR (1986): Psychosocial stress and behavior in primates. In Mitchell G, Erwin J (eds): "Comparative Primate Biology, Vol. 2A: Behavior, Conservation, and Ecology." New York: Alan R. Liss, Inc., pp 455–492.

Kaplan JR, Zucker E (1980): Social organization in a group of free ranging patas monkeys. Folia Primatol 34:196–213.

Kaplan JR, Nicks E, Riley J (1985): Behavioral responses of patas monkeys to group introductions. Am J Phys Anthropol 66:187.

Kaufmann JH (1967): Social relations of adult males in a free ranging band of rhesus monkeys. In Altmann SA (ed): "Social Communication Among Primates." Chicago: Chicago University Press, pp 73–98.

Kawai N (1958): On the system of social ranks in a natural group of Japanese monkeys. Primates 1:111–148.

Krige PD, Lucas JW (1975): Behavioral development of the infant vervet monkey in a free-ranging troop. J Behav Sci 2:151–160.

Kummer H (1967): Tripartite relations in hamadryas baboons. In Altmann SA (ed):"Social Communication Among Primates." Chicago: University of Chicago Press,pp 63–71.

Kummer H (1974): Distribution of interindividual distances in patas monkeys and gelada baboons: Species and sex differences. Folia Primatol 21:153–160.

Kurland J (1977): Kin selection in the Japanese monkey. Contrib Primatol 11.

Lancaster JB (1971): Play-mothering: The relations between juvenile females and young infants among free-ranging vervet monkeys (*Cercopithecus aethiops*). Folia Primatol 15:161–182.

Mason W (1985): Temperament and physiology: Ontogenetic and phylogenetic aspects. Am J Primatol 8:351.

Massey A (1977): Agonistic aids and kinship in a group of pigtailed macaques. Behav Ecol Sociobiol 2:31–40.

Mendoza SP, Moberg GP (1985): Species differences in adrenocortical activity of New World Primates: Response to dexamethasone suppression. Am J Primatol 8:215–224.

Oswald M, Erwin J (1976): Control of intragroup aggression by male pigtailed monkeys (*Macaca nemestrina*). Nature 262:686–687.

Oswald M, Lockard JS (1980): Ethogram of the de Brazza's guenon (*Cercopithecus neglectus*) in captivity. Appl Anim Ethol 6:285–296.

Rowell TE (1966): Hierarchy in the organization of a captive baboon group. Anim Behav 14:430–443.

Rowell TE (1971): Organization of caged groups of *Cercopithecus* monkeys. Anim Behav 19:625–645.

Rowell TE (1973): Social organization of wild talapoin monkeys. Am J Phys Anthropol 38:593–598.

Rowell TE (1974): The concept of social dominance. Behav Biol 11:131–154.

Rowell TE, Richards SM (1979): Reproductive strategies of some African monkeys. J Mammalogy 60:58–69.

Sade DS (1965): Some aspects of parent-offspring relations in a group of rhesus monkeys, with a discussion of grooming. Am J Phys Anthropol 23:1–18.

Sade DS (1967): Determinants of dominance in a group of free-ranging rhesus monkeys. In Altmann SA (ed): "Social Communication Among Primates." Chicago: University of Chicago Press, pp 99–114.

Sade DS (1972): A longitudinal study of social behavior of rhesus monkeys. In Tuttle R (ed.): "The Functional and Evolutionary Biology of Primates." Chicago: Aldine-Atherton, pp 378–398.

Sade DS (1980): Population biology of free-ranging rhesus monkeys on Cayo Santiago, Puerto Rico. In Cohen MN, Malpass RS, Klein HG, (eds): "Biosocial Mechanisms of Population Regulation." New Haven: Yale University Press, pp 171–187.

Samuels A, Henrickson RV (1983): Outbreak of severe aggression in a captive group of *Macaca mulatta*. Am J Primatol 5:277–281.

Sapolsky RM (1982): The endocrine stress response and social status in the wild baboon. Horm Behav 16:279–292.

Schjelderup-Ebbe T (1935): Social behavior of birds. In Murchison C (ed): "Handbook of Social Psychology." Worster, MA., Clark University Press.

Schjelderup-Ebbe T (1977): Contributions to the social psychology of chickens. In Schein MW (ed): Social Hierarchy and Dominance.'' Stroudsburg, PA: Dowden, Hutchinson and Ross, Inc, pp 35–49.

Seay B, Schlottmann R, Thorne BM (1970): Maternal and filial behavior in monkeys. Dev Psychol 3:66–72.

Seyfarth RM (1980): The distribution of grooming and related behavior among adult female vervet monkeys. Anim Behav 28:798–813.

Struhsaker TT (1967a): Behavior of vervet monkeys and other Cercopithecines. Science 156:1197–1203.

Struhsaker TT (1967b): Social structure among vervet monkeys (*Cercopithecus aethiops*). Behaviour 29:83–121.

Struhsaker TT (1969): Correlates of ecology and social organization among some African Cercopithecines. Folia Primatol 11:80–118.

Struhsaker TT (1971): Social behavior of mother and infant vervet monkeys (*Cercopithecus aethiops*). Anim Behav 19:233–250.

Struhsaker TT (1977): Infanticide and social organization in the redtail monkey (*Cercopithecus ascanius schmidtti*) in the Kibale Forest, Uganda. Z Tierpsychol 45:75–84.

Struhsaker TT, Gartlan JS (1970): Observations on the behavior and ecology of the patas monkey (*Erythrocebus patas*) in the Waza Reserve, Cameroon. J Zoo 161:49–63.

Struhsaker TT, Leland L (1979): Socioecology of five sympatric monkey species in the Kibale Forest, Uganda. Adv Study Behav 9:159–228.

Strum SC (1982): Agonistic dominance in male baboons: An alternative view. Int J Primatol 3:175–202.

Tokuda K, Jensen G (1968): The leaders role in controlling aggressive behavior. Primates 9:319–322.

Wade TD (1978): Status and hierarchy in nonhuman primate societies. In Bateson PPG, Klopfer PH (eds): ''Perspectives in Ethology.'' London: Plenum Press, pp 109–134.

Wilson EO (1975): ''Sociobiology: The New Synthesis.'' Cambridge, Massachusetts: Belknap-Harvard.

Wilson ME, Gordon TP, Bernstein IS (1978): Timing of births and reproductive success in rhesus monkey social groups. J Med Primatol 7:202–212.

Wolfheim JH (1977): A quantitative analysis of the organization of a group of captive talapoin monkeys (*Miopithecus talapoin*). Folia Primatol 27:1–27.

Wolfheim JH, Rowell TE (1972): Communication among captive talapoin monkeys (*Miopithecus talapoin*). Folia Primatol 18:224–255.

Wrangham RW (1980): An ecological model of female bonded primate groups. Behaviour 75:262–300.

Wrangham RW (1981): Drinking competition in vervet monkeys. Anim Behav 29:904–910.

Zucker EL, Kaplan JR (1981): A reinterpretation of ''sexual harassment'' in patas monkeys. Anim Behav 29:957–958.

Comparative Behavior of African Monkeys, pages 151–173

8

Social Status and the Distribution of Social Behavior by Adult Female Patas Monkeys: A Comparative Perspective

Evan L. Zucker

Department of Psychology, Loyola University, New Orleans, Louisiana 70118

INTRODUCTION

"Who does what to whom" is one of the most frequently mentioned phrases by those who study the social interactions of nonhuman primates. Whereas the "whos" (initiators) and the "whats" (behaviors) have been studied intensively and extensively, the "whoms" (recipients) have not received as much empirical attention. In those cases where the recipients have been addressed, the approach frequently taken has been to examine whether some individual or age class initiates some behavior more than it receives that behavior. Recipients are frequently described in terms of their age and sex classes, on the basis of whether or not they are kin or nonkin, or whether they are dominant or subordinate to the initiator. Thus, recipients are usually categorized dichotomously (male/female, adult/immature, kin/nonkin, dominant/subordinate), whether those categories have clear biologically-based boundaries or not.

Recipients of behaviors have been studied by those who have investigated the organizing role of rank-related attractiveness among adult females [Seyfarth, 1976, 1977, 1980; Fairbanks, 1980] and males [Colvin, 1983], where the dominance ranks of those individuals receiving the behaviors (grooming and agonistic aids) were critical in the development of the model. Also, in a recent series of papers, Bernstein and Ehardt [1985a, b, 1986] devoted more attention to the recipients of intragroup aggression and agonistic aids. The recipients were described using several dimensions, such as age, sex, and genealogical relationship. The inclusion of multiple dimensions provided increased detail about the relationships between the "whos" and the "whoms," although differences in status between the initiators and recipients were not reported.

Rowell [1966, 1971] described the relationships between the "whos" and "whoms" by indicating what percentage of some social behavior (e.g., grooming) was directed "up the hierarchy" or "down the hierarchy." While this measure provides a quantitative index of directionality, it does not address how far up the hierarchy, how far down, or which individuals within the hierarchy are doing the grooming. An individual is limited by its own position in the hierarchy. This is not to imply that animals know their positions [see Seyfarth, 1981], but that a third ranking individual, for example, can only groom two others if grooming up the hierarchy is to occur.

The primary focus of this chapter is whether an individual's position in a dominance hierarchy, constructed from outcomes of agonistic interactions, influences the organization of other social behaviors. With a hierarchy based on the outcomes of agonistic interactions, the top-ranked (alpha) individual would be "dominant" in all of its agonistic relationships, and conversely, the bottom-ranked (omega) individual would be "subordinate" in all of its agonistic relationships. Following Bernstein's [1981] view of dominance being defined socially rather than as a characteristic of an individual (p. 422), then all other group members will be in some dyadic social contexts as the more dominant member, but will also be in some interactions as the subordinate member. Even the second-ranked (beta) monkey would act as a subordinate when interacting with the alpha individual, and one of the two bottom-ranked individuals would be dominant to the other in their dyadic interactions. As most social groups of Old World monkeys are larger in size than only two individuals, then most of the group members participate in some interactions as the dominant individual, while in other contexts, this same individual participates as the subordinate individual. This also can be extended to genealogies, where there is a rank ordering of the genealogies within a social group.

An individual that loses most of its agonistic interactions is considered low ranking; are individuals such as these thereby limited with respect to social interactions and social partners? Does the capacity to show behavior appropriate for both dominant and subordinate roles, depending upon the social partner, have any bearing on whom this individual interacts with in other, nonantagonistic social contexts? What are the status differentials between the "whos" and the "whoms?" These questions are addressed in this chapter, using the free-ranging group of patas monkeys at La Parguera, Puerto Rico, as the focal point. The patterns of behavior of the adult females of this group will be compared to the corresponding patterns for several other species of Old World monkeys. When the hierarchical organization of the group members is known, then the position of "whom," relative to "who," can be taken into account in comparative investigations of the organization of social behavior.

METHODS

Subjects

Patas monkeys. Aspects of the social behavior of the patas monkeys (*Erythrocebus patas*) who lived on the island of El Guayacan, La Parguera, Puerto Rico, form the basis for comparisons. There were 18 adult females in this group, as well as one adult male, 14–15 infants, and 22 other immature monkeys. Additional information about this group and the island habitat are available elsewhere [Kaplan and Zucker, 1980; Vandenbergh, 1967; Zucker, this volume; Zucker and Kaplan, 1981a, 1981b].

Rhesus monkeys. Rhesus monkeys (*Macaca mulatta*) are included for comparative purposes. Data from one of 14 groups housed in 7.6 m × 7.6 m outdoor corrals at the Sabana Seca, Puerto Rico, facility of the Caribbean Primate Research Center are included [Kaplan et al, 1980; Kessler et al, 1985; Zucker, 1985]. This particular group was selected for inclusion because it contained 18 adult females, making it comparable numerically with the patas. This group, designated as group 30, contained one adult male. Composite data, based on all 14 groups, are included for comparative purposes where appropriate.

Allen's swamp monkey/Sykes' monkey. Animals in a mixed-species exhibit at the Central Florida Zoological Park are also included for comparative purposes. An Allen's swamp monkey male (*Cercopithecus nigroviridus*) was housed with a Sykes' monkey female (*C. mitis*). This pair was observed in the outdoor portion of the indoor/outdoor exhibit, which contained a cement water trough, wooden branches that provided arboreal access, elevated benches, and a tree stump on the cement floor. Background information on these two monkeys was not available.

Data Collection

Data describing social grooming and intergender agonism were obtained for all three groups. Additionally, data describing the distribution of allomaternal behaviors were available for the patas [Zucker and Kaplan,

1981a]. Hours of observation, and types of data collected, are as follows: For patas, 395.9 hr of ad libitum samples [Kaplan and Zucker, 1980]; for rhesus, 99.7 hr of 5-min focal samples, supplemented with ad libitum observations, on the 14 groups, including 7.33 hr on group 30; and for the swamp/Sykes' pair, 11.2 hr of 15-min focal samples. Patas data were collected between June 1977 and February 1978, rhesus data between November 1978 and May 1979, and the swamp/Sykes' data between October and December 1982.

Data Analyses

For all the comparisons included in this chapter, whether the source of data was from direct observations or the literature, the dominance relationships (hierarchies) among females needed to be known. These hierarchies were then subdivided into thirds (or "segments") for data analysis purposes. The thirds were designated as high, middle, and low. The frequencies of the target behavior (i.e., grooming, allomaternal acts, and intergender agonism) initiated by each third of the hierarchy were determined. These frequencies were tested for differences using chi-square goodness-of-fit tests. The data for each third were partitioned further, and the frequencies of initiating the target behavior as a function of the status of the recipients were determined. Thus, for example, the number of grooming bouts initiated by high-ranking females was partitioned according to whether the recipient was another high-ranking female, a middle-ranking female, or a low-ranking female. These "distributions of behavior" were tested for differences using chi-square goodness-of-fit tests. The null hypotheses for these tests were that behaviors within groups were initiated in equal frequencies by each third of the hierarchy, or that behavior was distributed independently of the rank of the recipients. The number of possible recipients within each third was used in determining the expected values for the chi-square tests. For all chi-square tests, the number of degrees of freedom was 2. For each of the behaviors considered, the patas data are presented first, followed by the comparative data.

RESULTS

Grooming Among Females: Initiators

Patas. The patas group of La Parguera was characterized by the low ranking females accounting for nearly one half of the grooming observed between adult females. High-ranking females initiated 15% of the grooming, middle-ranking females 41%, and low-ranking females 44% (X^2 = 17.22,

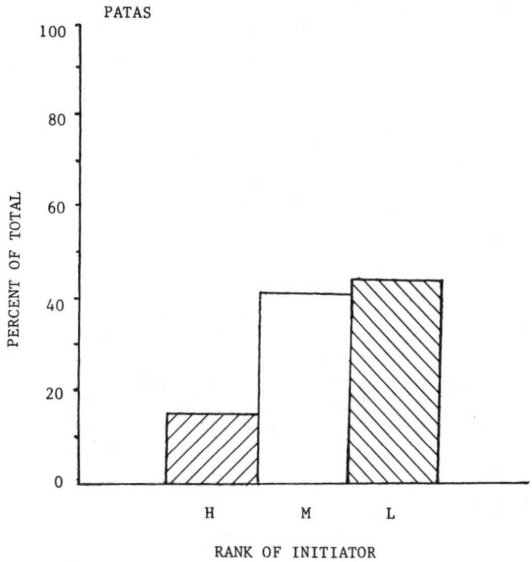

Fig. 1. The percentages of grooming initiated by adult female patas monkeys as a function of their status within the group. There were six females in each third of the hierarchy. H, M, L: high-, middle-, and low-ranking.

$n = 113$, $p < .001$; Fig. 1). For the patas group, 51% of the grooming was directed "up the hierarchy."

Comparative data. The percentages of grooming initiated by each third of the female hierarchy for several species of Old World monkeys are presented in Figure 2. For feral vervets, the composite pattern from three groups is presented, and for rhesus, composite data from 14 captive groups are presented.

For the feral vervets [data from Seyfarth, 1980], the high-ranking females initiated the most grooming (49% of the total), and the middle-ranking and low-ranking segments were approximately equal in their percentages of initiating grooming (24 and 27%, respectively). High-ranking females initiated significantly more grooming than was expected by chance, and middle-ranking and low-ranking females initiated less ($X^2 = 209.8$, $n = 2348$, $p < .001$). Two of the three vervet groups (Seyfarth's groups B and C) showed the same pattern as the composite; in group A, however, middle-ranked females were intermediate in their grooming frequencies, initiating less than the high-ranking females, but more than the low-ranking females.

Bonnet macaques [data from Koyama, 1973] initiated grooming in

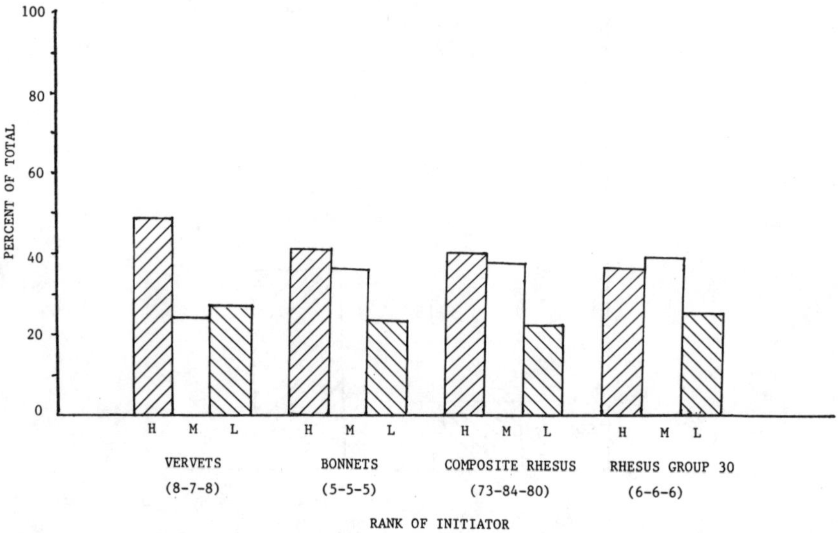

Fig. 2. The percentages of grooming initiated by adult female vervets, bonnets, and rhesus. The numbers under the names indicate the number of females in each segment of the hierarchy. H, M, L: high-, middle-, and low-ranking.

frequencies that reflected relative rank; high-ranking females initiated a higher percentage (41%) than did middle-ranking females (36%) and low-ranking females (23%) ($X^2 = 36.0$, $n = 767$, $p < .001$). High-ranking females initiated significantly more than expected by chance, and low-ranking females less than expected.

The composite pattern for the Sabana Seca rhesus monkeys closely resembles the pattern of the bonnet macaques. High-ranking females initiated 41% of the grooming, middle-ranking 37%, and low-ranking 22%. Like the bonnets, high-ranking females initiated more grooming than was expected by chance, and low-ranking females less than expected ($X^2 = 48.0$, $n = 699$, $p < .001$). For both bonnets and rhesus, middle-ranking females initiated grooming as expected based on their proportions in the groups.

For rhesus group 30, high-ranking females initiated 36% of the grooming, middle-ranking females 39%, and low-ranking females 25%. These frequencies for initiating grooming were not different from a pattern of equal frequencies ($X^2 = 1.16$, $n = 36$, $p > .05$).

These data from Old World monkey females were converted to the percentage of grooming that was directed "up the hierarchy." These percentages are presented in Table I, along with additional data from Rowell

TABLE I. Comparative Data for the Percentage of Grooming "Up the Hierarchy" for Six Species of Old World Monkey Females

Species	No. of females	No. of bouts	% "up the hierarchy"	Reference
Patas (*E. patas*)	18	113	51	Kaplan and Zucker [1980]
Vervets (*C. aethiops*)				
Group A	8	1,064	53	Seyfarth [1980]
Group B	7	325	47	Seyfarth [1980]
Group C	8	959	53	Seyfarth [1980]
Composite	23	2,348	52	Seyfarth [1980]
Vervets (*C. aethiops*)	6	1,743	58[a]	Rowell [1971]
Sykes' (*C. mitis*)	5	1,965	68[a]	Rowell [1971]
Baboons (*P. anubis*)	5	2,190[b]	39	Rowell [1966]
Bonnets (*M. radiata*)	15	767	56	Koyama [1973]
Rhesus (*M. mulatta*)				
Group 30	18	36	59	Zucker [1985]
Composite	237	699	52	Zucker [1985]

[a]Percentage includes grooming of an adult male.

[b]Frequency for all grooming within the group.

[1966, 1971] for vervets, Sykes' monkeys, and olive baboons. For the six species, the indices of directionality are in the 50–60% range. Rowell's [1971] finding of 68% for Sykes' monkeys and the 39% for olive baboons [Rowell, 1966] constitute the extreme high and low, respectively. In contrast to the patas' 51%, feral vervets groomed "up" 52% of the time, feral bonnets 56%, and captive rhesus 52%.

If only these percentages were used as a basis of comparisons, then the vervets and rhesus would be described as showing similar patterns of grooming as a function of hierarchical organization. The bonnets' grooming would be more strictly influenced by dominance relationships, and the patas would be least limited by dominance relationships. However, when the status of the recipients of grooming are taken into account, different patterns emerged.

Grooming Among Females: Recipients

Patas. When the relationship between initiator and recipient is taken into account (Fig. 3), it is evident that when the high-ranking patas females groomed other females, the recipients tended to be other high-ranking females ($X^2 = 44.4$, $n = 17$, $p < .001$). Middle-ranking females directed their grooming toward low-ranking females preferentially ($X^2 = 48.3$, $n = 46$, $p < .001$), whereas low-ranking females directed most of their grooming

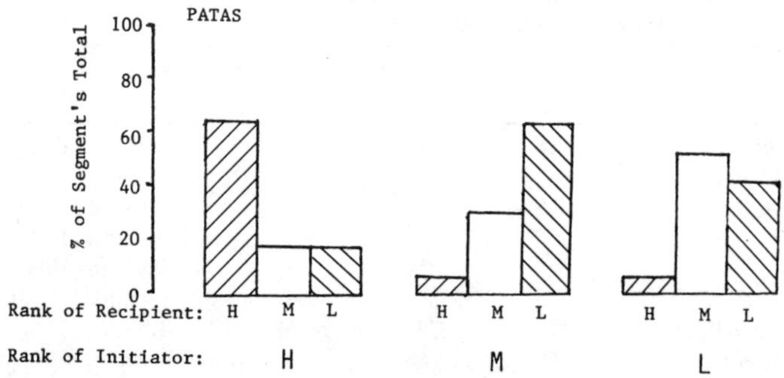

Fig. 3. The distribution of social grooming, expressed as the percentage of each segment's total grooming, for adult female patas monkeys. H, M, L: high-, middle-, and low-ranking.

to middle-ranking and other low-ranking females ($X^2 = 35.1$, n = 50, $p < .001$). For these latter two segments, very little grooming was directed towards the females in the high-ranking third of the group.

Comparative data. The distribution of grooming among vervet females resembled that of the patas, but only for the highest ranking third of the hierarchy (Fig. 4). The combined data for the three vervet groups [Seyfarth, 1980] showed that high-ranking females groomed other high-ranking vervet females significantly more than was expected by chance ($X^2 = 338.6$, n = 1151, $p < .001$). Middle-ranking vervets also groomed high-ranking vervets significantly more than was expected; they groomed similarly ranked individuals the least ($X^2 = 84.1$, n = 557, $p < .001$). Low-ranking vervets did not show any preference for the status of the recipients of their grooming ($X^2 = 3.59$, n = 639, $p > .05$).

Whereas vervets and rhesus both distributed 52% of their grooming up the hierarchy, the distribution of grooming for rhesus, based on the status differences between the initiators and the recipients, distinguishes these two species. In comparison to the patterns described above, high-ranking female rhesus did not groom other high-ranking females as exclusively as did either patas or vervets (Fig. 4); high-ranking rhesus females groomed middle-ranking females more than did females of the other two species. Middle-ranking rhesus females groomed other high-ranking females most frequently, but they groomed similarly ranked individuals more than did their vervet counterparts. Low-ranking rhesus females groomed middle-ranking females the most, and high-ranking females the least. Patterns for all three segments were significantly different from equal distributions ($p < .001$ each third).

Bonnet macaques showed still a different pattern. Low-ranking bonnet

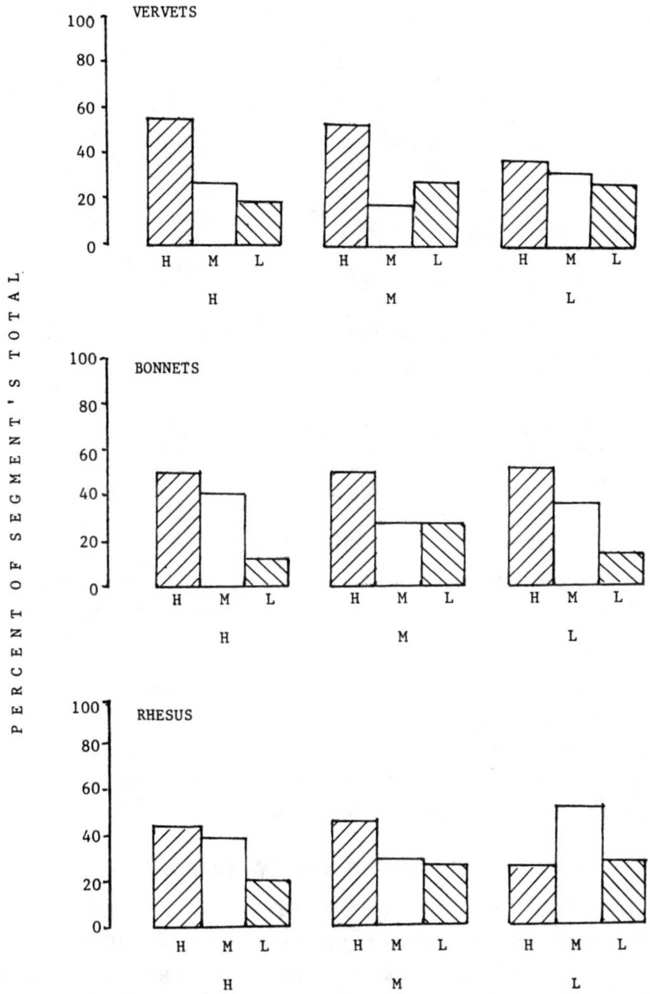

Fig. 4. The distribution of social grooming, expressed as the percentage of each segment's total grooming, for adult female vervets, bonnets, and rhesus. H, M, L: high-, middle-, and low-ranking.

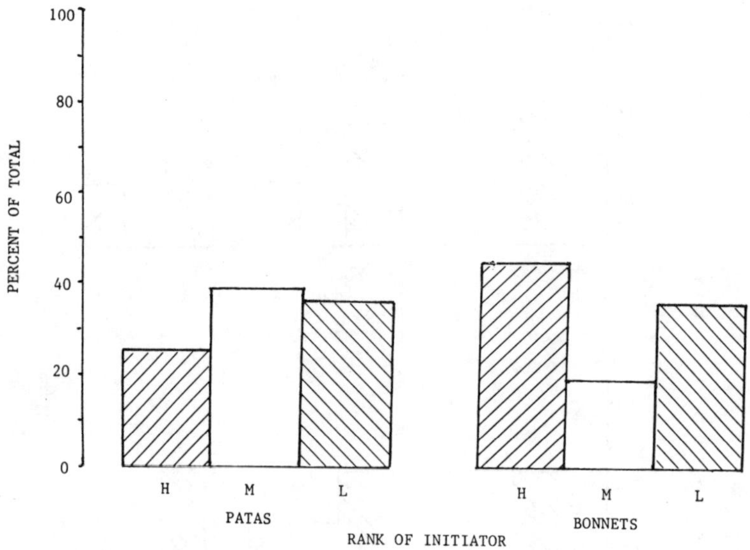

Fig. 5. The percentage of allomaternal behaviors initiated by adult female patas and bonnet monkeys as a function of the status of the initiators. For the patas, there were six females in each portion of the hierarchy; for the bonnets, the thirds of the hierarchy contained nine, ten, and nine females, respectively. H, M, L: high-, middle-, and low-ranking.

females received very little grooming (Fig. 4). When they were groomed, it was most frequently from middle-ranking females (e.g., "down the hierarchy"). High-ranking bonnet females received the largest proportion of grooming regardless of the initiator's status.

Allomaternal Behaviors: Initiators and Recipients

Comparative data for allomaternal behaviors are limited. Silk's [1980] data for kidnapping and attempted kidnapping in a group of bonnet monkeys living in an outdoor field cage are compared to the patterns of allomaternal behavior described for the free-ranging patas monkeys [Zucker and Kaplan, 1981a]. In these comparisons, the status of the infants' mothers was used in determining the recipients' status [see Loy and Loy, this volume, regarding ranks of mothers and their offspring].

The initiators of allomaternal behavior for the patas and bonnets are shown in Figure 5. For patas, there was no significant difference across thirds of the hierarchy for initiating an allomaternal interaction ($X^2 = 3.0$, $n = 92$, $p > .05$). For bonnets, there was a significant deviation from equal frequencies of initiations according to status ($X^2 = 27.1$, $n = 196$, $p < .001$). High-

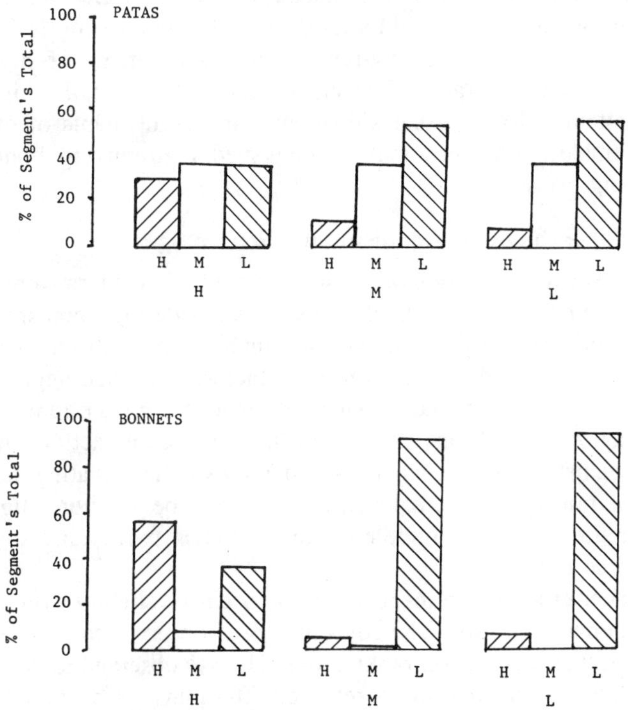

Fig. 6. The distribution of allomaternal interactions, expressed as the percentage of the total for each segment, for adult female patas and bonnet monkeys. The number of females in each segment is as described for Figure 5. H, M, L: high-, middle-, and low-ranking.

ranking bonnet females were responsible for more kidnappings and attempted kidnappings than was expected, and middle-ranking females did less than expected.

When the recipients of allomaternal behaviors are taken into account, several similarities emerged between the two species, but several differences also became evident (Fig. 6). High-ranking patas females distributed their allomaternal behaviors without regard for the ranks of the recipients ($X^2 = 0.4$, $n = 23$, $p > .05$), whereas the middle-ranking and low-ranking patas females showed nearly identical patterns. Both of these latter two segments directed more allomaternal behaviors toward infants of low ranked females, and directed few allomaternal behaviors of any type toward infants of high-ranking females.

Like the patas, middle-ranking and low-ranking bonnet monkeys showed nearly identical patterns. For bonnets, however, nearly all of the allomaternal behaviors of middle-ranking and low-ranking females were directed toward infants of low-ranking females; infants of high-ranking females were rarely targets. Conversely, most (80%) of the high-ranking females' attempted and actual kidnappings involved infants of other high-ranking females, a pattern that distinguished high-ranking bonnets from high-ranking patas.

Intergender Behaviors: Agonism and Grooming

For the patas of La Parguera, these intergender behaviors were relatively infrequent, precluding statistical analysis. Considering agonism first, the adult male initiated only seven agonistic interactions with females, and the rank of the female did not seem to be a factor. The male aggressed twice against a high-ranking female, four times against a middle-ranking female, and once against a low-ranking female. In agonistic interactions initiated by females against the male, where the identity of the initiating female was known, the ranks of the females did not seem to be a factor. High-ranking females aggressed once, middle-ranking females twice, and low-ranking females three times.

Intergender grooming in this patas group occurred slightly more frequently than intergender agonism. The adult male was observed to groom only one female, and that was the top-ranking one. He was observed to do this twice. On the other hand, the male received grooming from females on 26 occasions; of these, 25 (96%) were from females in the top third of the hierarchy [Kaplan and Zucker, 1980; see Kaplan, this volume, for a further discussion of patas grooming].

The swamp/Sykes' pair showed a pattern of agonism that was similar to that of the patas, in that the female Sykes' would retaliate against the male. Twenty-one agonistic encounters were observed for this pair. The male initiated 20 of these (95%), although in seven interactions, the female emitted some retaliatory behavior.

The pattern of grooming observed between members of this pair was similar to that seen for patas, as well as for many other species of Old World monkeys. In this mixed-species group, the female initiated 79% of the grooming (49/62 bouts), and groomed the male swamp monkey for a longer period of time per bout than the male groomed the female Sykes' monkey. Of course, with this pair, no considerations of rank were possible.

Comparative data for grooming were available for two species of macaques (bonnets and rhesus) where ranks of the females were known. High-ranking adult female bonnet macaques were responsible for 63% of the grooming directed toward males in the feral group studied by Koyama [1973]. The middle-ranking females of this group did 23% of the intergender grooming, and low-ranking females did 14% (n = 160). Grooming from males to females did not mirror this pattern. Males groomed middle-ranking females the most (50% of the 92 bouts), whereas they groomed high-ranking females 42% of the total; the low-ranking females received 8% of the total. For all the grooming reported for this bonnet group, female to male grooming accounted for 14% of the total, and male to female grooming accounted for 8%.

For the rhesus at Sabana Seca, males directed significantly more grooming toward high-ranking females than was expected by chance, and less towards low ranking females ($X^2 = 15.3$, $n = 37$, $p < .001$). When the females groomed males, high-ranking females accounted for approximately 60% of the total, middle-ranking females 32%, and low-ranking females 6% (n = 142). Thus, the grooming pattern, according to relative status, that was shown by rhesus females toward rhesus males was similar to that of bonnets. In comparing the percentages of the total amount of grooming within the groups, the rhesus and bonnets were quite similar. Female to male grooming comprised 16% of the total amount of grooming, and male to female grooming accounted for 4% of the total.

Data for intergender agonism were available for the rhesus groups at Sabana Seca. With respect to this intergender agonism, the rhesus and patas did not show similar patterns. Whereas the patas male received agonism from all thirds of the hierarchy, rhesus males received 86% (25/29) of their intergender agonism from high-ranking rhesus females. The remaining agonism came from middle-ranking females. Rhesus males directed more of their agonism toward high-ranking females (40% of the 127 interactions) than towards middle-ranking (38%) and low-ranking (22%) females. High-ranking females received more agonism than was expected by chance, and low-ranking females less than expected ($X^2 = 6.9$, $n = 127$, $p < .05$).

DISCUSSION

The cross-species comparisons presented illustrate some striking differences in the interaction patterns among the adult females of these groups, although similarities, too, were evident in the distribution of social behaviors. The females in each of the groups considered (except for the swamp/Sykes' pair) could be arranged in a linear dominance hierarchy that was based on outcomes of competitions [Koyama, 1973] or agonistic interactions [Seyfarth, 1980; Silk, 1980; Zucker, 1985; Zucker and Kaplan, 1980]. The influence that this hierarchy had on the organization of other social behaviors, however, was variable, not only in terms of which females initiated the particular social behavior, but in terms of with which other females they interacted.

Distribution of Grooming and Allomaternal Behavior

For the patas group, the "focal group" in these analyses, the low-ranking females were the ones that initiated the most grooming, and this third of the group, in conjunction with the middle-ranking females, were most involved in allomaternal behaviors. Yet, when the behaviors of the high-ranking females were examined in terms of the recipients of their nonagonistic social interactions, it was clear that these high-ranking females preferred other females of similar rank for both grooming and allomaternal behaviors. Thus, it appears that these females were more selective in their social interactions than were the females of other segments of the group.

High-ranking vervet females, like their patas counterparts, also preferred like-ranked females as grooming partners. Where vervets differed from patas was in the distribution of behavior by females in the lower two thirds of the hierarchies. Middle-ranking vervets preferred the higher-ranking females of their groups, whereas the low-ranking females did not preferentially groom any particular portion of the hierarchy.

In contrast, the Asian macaque females (bonnets and rhesus) showed similar patterns in their distributions of grooming, but only for the high-ranking and middle-ranking segments of these groups. For these two species, high-ranking females were the preferred partners. Low-ranking bonnet females also showed this preference, whereas low-ranking rhesus females demonstrated a preference for the middle-ranking females within their groups.

These different within-group patterns are interesting in light of the similar estimates of grooming "up the hierarchy" that were calculated or reported for these four species (see Table I). These differences in the distribution of grooming indicate that the initiators of grooming were constrained, in some respects, as to whom they directed their grooming, and conversely, from whom they received their grooming. Although certain females within groups were constrained, they were not necessarily limited in terms of the frequencies of interactions.

This constraint, associated with a female's relative position in a status hierarchy, was also evident in the distribution of allomaternal behaviors, although the availability of comparative data was limited. The data for the female bonnet monkeys [from Silk, 1980] contrast greatly with those data for the female patas. For the bonnets, high rank was associated with relative immunity from the attempted kidnappings of one's infant, except from other high-ranking females. Infants of low-ranking females were most susceptible to kidnapping attempts from both middle-ranking and other low-ranking females. The infants of middle-ranking females, however, were rarely targeted. For the patas, high-rank was associated with being able to interact with any female's infant, as no preferences were shown based on the ranks of the recipients. The infants of low-ranking patas females, on the other hand, were the most susceptible to allomaternal acts, but not to the same extent as their bonnet counterparts. Where the infants of low-ranking patas received just over half of the allomaternal acts from both middle-ranking and other low-ranking patas females, the corresponding percentages for the bonnets were 92 and 94%, respectively.

Silk's [1980] data for the bonnet group dealt only with kidnappings and attempted kidnappings, behaviors that were potentially detrimental to the infants, as well as detrimental to the reproductive successes of the infants' mothers [see Hrdy, 1976]. Whereas the infants of low-ranking females were at risk, infants of high-ranking females were also at risk, to some extent, as these infants were subjected to kidnapping attempts by females of similar rank as their mothers. Even the infant of the top-ranking female of the bonnet group was subjected to two attempted kidnappings [Silk, 1980, her Table 2]. Female bonnet monkeys may have disproportionately directed their kidnapping attempts toward subordinate individuals [Silk, 1980], but females of high rank directed their kidnapping attempts to those fairly close to their own status. Similarly, the greatest risk to the infants of low-ranking females was posed primarily by females who were also low-ranking.

A different situation, in terms of risks, existed for the patas. Allomaternal interactions in this group were primarily affiliative, including physical contact, grooming, and close visual inspection of the infants of other females [Zucker and Kaplan, 1981a]. Although several kidnappings occurred, mothers had little difficulty regaining their infants. Only two agonistic behaviors were observed directed toward infants by adult females. In this group, allomaternal interactions were more likely to be beneficial than were the interactions in the bonnet group, and lactating patas females perhaps were less protective of their infants upon the approach of another female of any rank. Infants of high-ranking mothers received few allomaternal acts from females ranked in the lower two thirds of the hierarchy, but, comparatively, they did receive a higher percentage from these females than did the bonnet macaques. As with social grooming, the differences in relative status constrained the interactions, but the status differences were not barriers to interactions between females of disparate ranks (and their infants).

Intergender Interactions

The extent of the nonsexual, intergender interactions, as well as the distribution of these interactions according to the relative ranks of the participants, provides indications of the degree of social integration of the adult males and females of a group. The available data indicate species differences; the bonnet and rhesus macaques were more similar to each other in their patterning of social interactions than the patas were similar to either of these two species.

Female patas, independent of their status within the group, were fairly equally involved in agonistic interactions with the male, both as the recipients of agonism and as the initiators. However, affiliative social interactions with the male were nearly exclusively within the domain of the high-ranking females.

The composite pattern for rhesus monkeys, for whom data for both agonistic and affiliative interactions were available, indicated that the high-ranking adult females were responsible for most of the agonism that was directed toward males. Whereas high-ranking rhesus females received more agonism than expected by chance, they did not recieve appreciably more than the middle-ranking females (40 vs 38%, respectively).

Intergender grooming predominantly involved the high-ranking female rhesus, although not to the same extent as in the patas group. The prevalence of participation by the high-ranking rhesus females in intergender grooming applied to both initiating and receiving grooming.

The bonnet females' grooming patterns, as initiators of intergender grooming, resembled that of the rhesus females. As recipients, however, middle-ranking females were preferred by the bonnet males. This apparent species difference might be an artifact of the compositions of the bonnet and rhesus groups, in that the bonnet group studied by Koyama [1973] contained nine males, whereas most (12/14) of the groups at Sabana Seca contained only one male. Inspection of Koyama's grooming matrix, however, shows that the higher-ranking males preferentially groomed the higher-ranking females (both high-ranking and mid-ranking). The frequencies of males receiving grooming from females indicate the presence of specific partner preferences. Only the top-ranking male received grooming from each of the five high-ranking females.

Chapais [1986] studied grooming relationships between males and females in group F on Cayo Santiago, a group with 15 males and 19 females. The top-ranking female received more grooming from the top-ranking three males than she gave to them. However, all other females showed preferences for these three top-ranking males. These grooming relationships indicate that the patterns observed in the Sabana Seca rhesus were not distorted by the relative unavailability of adult males. The lower-ranking males in group F on Cayo Santiago were infrequently involved in grooming with adult females [''low frequency'' or ''very occasional''; Chapais, 1986, Table 8.3, p. 185].

The data describing the interactions between the Allen's swamp monkey male and the Sykes' monkey female, albeit limited, are interesting in comparative perspective. The data for the swamp monkey were included, since behavioral data for this species are practically nonexistent [see Erwin and Zucker, this volume]. The pattern of grooming that was observed for this dyad was much like that of other Old World monkeys, despite the species difference and the difference in sizes; swamp monkeys are the largest of the *Cercopithecus* species [Napier and Napier, 1967]. The female initiated more grooming, and groomed the male for longer periods of time (per bout) than the male did the Sykes' female. Agonistically, the smaller Sykes' monkey was observed to initiate only one interaction, but she retaliated against male-initiated agonism.

The degree of sexual dimorphism might influence the expression of female-initiated intergender agonism in *Cercopithecus* (and *Miopithecus*) species. Vervet females did not aggress against adult male vervets, unless the females were part of a coalition [Struhsaker, 1967]. Seventy percent of all female-female coalitions among vervets were formed to defeat adult males in agonistic encounters [Cheney, cited in Seyfarth, 1980]. For talapoins, a species where sexual dimorphism is greatly reduced, if present at all [Rowell, 1973; Wolfheim and Rowell, 1972], individual intergender agonism was apparently more common. Adult males received 30% of their total agonism from adult females, and conversely, adult females received 67% of their total from adult males [Gautier-Hion, 1970]. However, the social status of the females involved in this intergender agonism was not considered.

The data for vervets and talapoins suggest a relationship between the occurrence of intergender dyadic agonism and the degree of sexual dimorphism. However, this relationship is not supported when the data from the patas of La Parguera are considered. Intergender agonism was relatively rare among the patas, and the female's status did not appear to be a factor. Female-initiated agonism against the male occurred despite the sexual dimorphism. Whereas sexual dimorphism might still be a contributing factor in intergender agonism, the male's degree of social involvement in the day-to-day activities of the group might also be a factor. The patas male tends to be spatially and socially aloof [e.g., Hall, 1965]. When near the group, he is vulnerable to aggression from the females. Male talapoins do not interact to a great extent with adult females during the nonbreeding season [Rowell, 1973; Rowell and Dixson, 1975]. Thus, when the males do become integrated into the group, they too might be more vulnerable to female-initiated aggression, perhaps from any female in the group.

In the above comparison between patas and talapoins, the variable of reproductive seasonality was introduced. The data presented in this chapter for intergender interactions were not partitioned to take this variable into account. Patas and rhesus, the two species for which the most detailed data were presented, are both seasonally breeding species [e.g. Drickamer, 1974; Varley and Vessey, 1977; Walker et al, 1983; Wilson et al, 1978, for rhesus], although the patas' breeding and birth seasons may not be as restrictive or rigid as that of rhesus [e.g., Rowell and Hartwell, 1978]. Whereas a patas female's reproductive state influences her male-directed behavior [Loy, 1981], the effect of reproductive state on the behavior of females at La Parguera was likely minimal. At the time of the observations of this group, 15 of the 18 females had suckling infants, indicative of seasonal breeding. The females without infants were not clustered in any one segment of the hierarchy, so any differences between females in the various segments of the group were unlikely due to any differences in reproductive states. Differences in the behavior of the female rhesus as a function of their reproductive state could not be assessed. As the behavior of 237 females was being monitored, there were insufficient data on individual females to judge any effects of reproductive state. Additionally, at the time of these observations, a management strategy of minimal handling and intervention was imposed [Kaplan et al, 1980; Kessler et al, 1985]. Thus, techniques to assess reproductive states, such as vaginal swabbing or venipuncture, were not possible. In the absence of this information, the reproductive states described for the La Parguera patas should also apply to the rhesus. The greater the degree of reproductive synchrony among females, the less of an effect changes in reproductive states should have in the distribution of behavior as a function of relative status. If all or most of the females in a group are either receptive, pregnant, or lactating at the same time, the effects of reproductive state should be balanced across portions of a hierarchy (see Chapais, 1983, 1986, for more information about rhesus intergender grooming and reproductive states).

Dominance and the Distribution of Behavior

The analyses presented in this chapter were intended to address several questions about the role of dominance in the organization of social behavior for several species of Old World monkeys. The questions followed from several of Bernstein's [1981, 1984] statements regarding dominance. According to Bernstein [1981]:

1. Agonistic dominance ranks need not influence all social interactions (p. 427).
2. Agonistic rank hierarchies must be shown to act in their own right as an independent variable influencing one or more dependent variables before it can be concluded that agonistic ranks are an important feature of social organization (p. 428).
3. We have little evidence that any specific rank causes an individual to behave in a specific fashion, or that a knowledge of numerical ranks allows us to accurately predict the behavior of specific individuals (p. 428).

These statements, treated as hypotheses, formed the framework for the comparative analyses presented in this chapter.

With respect to the pervasive influence of agonistic dominance relationships, the grooming data for the females of all four species (patas, vervets, bonnets, and rhesus) indicated that all portions of the hierarchy interacted to some extent with all other portions of the hierarchy. Thus, agonistic dominance ranks did not influence all grooming interactions. For allomaternal behaviors, female patas monkeys interacted with the infants of other females regardless of the relative status of the infants' mothers. Whereas there were preferences for interactions with mothers and infants of particular segments of the hierarchy, agonistic dominance relationships and status did not preclude interactions with any other segment. Dominance relationships among bonnets, however, were more restrictive, at least for the bonnet group studied by Silk [1980]. In this group, low-ranking females did not attempt any kidnappings of infants of the middle-ranking females. It is important to note, however, that of the middle-ranking females in the group, only two had infants during that particular birth season. Whereas the females in all of the other segments interacted with each other (and each other's infants), this absence of interactions between low-ranking females and the infants of middle-ranking females is likely an artifact of the group composition.

Agonistic rank hierarchies served to organize grooming between adult females, but the extent of that organization varied with the segment of the group and the species. For example, agonistic rank seemed to influence the choice of grooming partners for high-ranking patas monkeys, who distributed 65% of all their grooming to other high-ranking females. Similarly, high-ranking bonnet females were the preferred recipients of grooming no matter what the relative status of the initiator. The lack of any organizing effect of rank relationships was also seen. Low-ranking vervet females distributed their grooming without preference for partners of particular ranks (e.g. thirds of the heirarchy). Middle-ranking and high-ranking vervet females did show preferences. Thus it appears that there is an interaction effect (in a statistical or predictive sense) between the portion of the hierarchy and the species.

Knowledge of an individual's specific rank may not permit the prediction of the exact pattern of that individual's interactions. However, it does appear that knowledge of relative rank, expressed as high, middle, or low in these analyses, may allow statements to be made regarding the probabilities of interactions with particular group members based on the relative status of those other group members. While it is true that the data presented here for the patas were derived from a single group and that the bonnet data came from two independent studies of different groups, knowledge of the patterns described here should aid in the formulation of testable hypotheses regarding the organization of social behavior, and should provide a basis for additional comparative work. Generalizations about the social behavior of nonhuman primates and the formulation of models of behavior will likely be more accurate when they are developed from comparative studies of multiple groups.

The data included here for vervet monkeys [from Seyfarth, 1980] indicated that behavior patterns across groups of a single species might not vary greatly if the ecological conditions are similar for the different groups. The patterns seen for the three groups of feral vervets that Seyfarth studied were similar, and the composite patterns were not very different from those of the individual groups. It is not really possible to determine whether or not the composite pattern for the Sabana Seca rhesus accurately reflected the patterns of the 14 groups, as the amount of data per group was not extensive. It was for this reason that the composite patterns were presented.

It is possible that some of the patterns for the distribution of affiliative behaviors reflected the organizing effects of genealogies rather than agonistic dominance ranks. Whereas this might be the case, the division of the hierarchies into thirds should increase the likelihood of including entire genealogies within any segment, while still being a unit of analysis that is small enough to allow for the detection of differences. If only the divisions of high-ranking and low-ranking were to be used, then it is more probable that entire genealogies will be included in the same division. However, the partitioning of large groups into halves is likely to prevent the detection of subtle differences in the patterns of social interaction.

In contrast to analyzing patterns of social interactions according to some partitioning of the hierarchy, Chapais [1983] described grooming by adult female rhesus monkeys according to individual patterns. Five different patterns were identified that described the grooming relationships of 17 females (ranked 3 through 19) in group F on Cayo Santiago. These patterns, however, only described the grooming that was directed "up the hierarchy," and partner preferences were evident that did not appear to be rank-related. As grooming can go either up or down a hierarchy, as well as different distances up or down, the efficacy of using these individual patterns, as well as any other way of partitioning groups, requires further evaluation and comparative analysis.

CONCLUSIONS

1. Low ranking patas females initiated the most grooming in a free-ranging group, whereas for vervets and bonnets, the high-ranking thirds of the respective hierarchies initiated the most grooming. Combined data for 14 captive rhesus groups indicated the same pattern as the vervets and bonnets, although for one group containing the same number of females as the patas group, the middle-ranking females initiated the most grooming.

2. When the relative status of the initiators and recipients of grooming were analyzed, different patterns were evident. Species differences in the distribution of grooming, by thirds of the hierarchy, were found, as were preferences for females of particular ranks as partners. These preferences, however, were dependent upon the status of the initiators.

3. High-ranking patas females engaged in allomaternal acts without regard for the ranks of the infants' mothers, whereas other patas females preferred to interact with the infants of the lower-ranking females. For bonnets, high-ranking females attempted kidnappings of infants of other high-ranking females. All other females directed their kidnapping attempts nearly exclusively toward infants of low-ranking females.

4. The amount and patterning of grooming and agonism between adult males and females varied with the species. The patterns of agonism might be related to the degree of sexual dimorphism and the degree of male integration into the daily social activities of the group.

5. The organization of social interactions was constrained by differences in ranks between initiators and recipients, but was not prevented by these rank differences.

6. The rank relationships between females, based on agonistic outcomes, did not influence all of their other social interactions. However, the degree

to which ranks influenced the distribution of affiliative social behaviors varied according to the species, and according to the relative ranks of the females within the group.

ACKNOWLEDGMENTS

I gratefully acknowledge the following individuals for their assistance and contributions to the research and ideas that were included here: J. R. Kaplan, M. J. Kessler, W. T. London, W. T. Kerber, H. J. Herbert, G. Meier, S. Ebbeson, H. S. Martinez, E. Posey, T. L. Maple, A. T. Pietrewicz, R. McCleary, and M. R. Clarke. Financial support for the observations of the free-ranging patas was provided by a faculty grant from the University of Alabama in Birmingham to J. R. Kaplan, and supplemental support was provided through the Psychology Department of Emory University; support for the work done with the rhesus at Sabana Seca was provided by W. T. London.

REFERENCES

Bernstein IS (1981): Dominance: The baby and the bathwater. Behav Brain Sci 4:419–429.

Bernstein IS (1984): The adaptive value of maladaptive behavior, or you've got to be stupid in order to be smart. Ethol Sociobiol 5:297–303.

Bernstein IS, Ehardt CL (1985a): Age-sex differences in the expression of agonistic behavior in rhesus monkey (*Macaca mulatta*) groups. J Comp Psychol 99:115–132.

Bernstein IS, Ehardt CL (1985b): Agonistic aiding: Kinship, rank, age, and sex influences. Am J Primatol 8:37–52.

Bernstein IS, Ehardt CL (1986): Selective interference in rhesus monkey (*Macaca mulatta*) intragroup agonistic episodes by age-sex class. J Comp Psychol 100:380–384.

Chapais B (1983): Dominance, relatedness and the structure of female relationships in rhesus monkeys. In Hinde RA (ed): "Primate Social Relationships." Sunderland: Sinauer Associates, pp 208–219.

Chapais B (1986): Why do adult males and female rhesus monkeys affiliate during the birth season? In Rawlins RG, Kessler MJ (eds): "The Cayo Santiago Macaques: History, Behavior, and Biology." Albany: State University of New York Press, pp 173–200.

Colvin J (1983): Familiarity, rank, and the structure of rhesus male peer networks. In Hinde RA (ed): "Primate Social Relationships." Sunderland: Sinauer Associates, pp 190–200.

Drickamer LC (1974): A ten-year summary of reproductive data for free-ranging *Macaca mulatta*. Folia Primatol 21:61–80.

Fairbanks LA (1980): Relationships among adult females in captive vervet monkeys: Testing a model of rank-related attractiveness. Anim Behav 28:853–859.

Gautier-Hion A (1970): L'organisation sociale d'une bande de talapoins (*Miopithecus talapoin*) dans le nord-est du Gabon. Folia Primatol 12:116–141.

Hall KRL (1965): Behaviour and ecology of the wild patas monkey, *Erythrocebus patas*, in Uganda. J Zool 148:15–87.

Hrdy SB (1976): Care and exploitation of nonhuman primate infants by conspecifics other than the mother. Adv Study Behav 6:101–158.

Kaplan JR, Zucker E (1980): Social organization in a group of free-ranging patas monkeys. Folia Primatol 34:196–213.

Kaplan JR, Manning P, Zucker E (1980): Reduction of mortality due to fighting in a colony of rhesus monkeys (*Macaca mulatta*). Lab Anim Sci 30:565–570.

Kessler MJ, London WT, Rawlins RG, Gonzalez J, Martinez HS, Sanchez J (1985): Management of a harem breeding colony of rhesus monkeys to reduce trauma-related morbidity and mortality. J Med Primatol 14:91–98.

Koyama N (1973): Dominance, grooming, and clasped-sleeping relationships among bonnet monkeys in India. Primates 14:225–244.

Loy J (1981): The reproductive and heterosexual behaviours of adult patas monkeys in captivity. Anim Behav 29:714–726.

Napier J, Napier P (1967): "A Handbook of Living Primates." New York: Academic Press.

Rowell TE (1966): Hierarchy in the organization of a captive baboon group. Anim Behav 14:430–443.

Rowell TE (1971): Organization of caged groups of *Cercopithecus* monkeys. Anim Behav 19:625–645.

Rowell TE (1973): Social organization of wild talapoin monkeys. Am J Phys Anthropol 38:593–598.

Rowell TE, Dixson AF (1975): Changes in social organization during the breeding season of wild talapoin monkeys. J Reprod Fertil 43:419–434.

Rowell TE, Hartwell KM (1978): The interaction of behavior and reproductive cycles in patas monkeys. Behav Biol 24:141–167.

Seyfarth RM (1976): Social relationships among adult female baboons. Anim Behav 24:917–938.

Seyfarth RM (1977): A model of social grooming among adult female monkeys. J Theor Biol 65:671–698.

Seyfarth RM (1980): The distribution of grooming and related behaviours among adult female vervet monkeys. Anim Behav 28:798–813.

Seyfarth RM (1981): Do monkeys rank each other? Behav Brain Sci 4:447–448.

Silk JB (1980): Kidnapping and female competition among bonnet macaques. Primates 21:100–110.

Struhsaker TT (1967): Social structure among vervet monkeys (*Cercopithecus aethiops*). Behaviour 29:83–121.

Vandenbergh JG (1967): The development of social structure in free-ranging rhesus monkeys. Behaviour 29:179–194.

Varley MA, Vessey SH (1977): Effects of geographic transfer on the timing of seasonal breeding of rhesus monkeys. Folia Primatol 28:52–59.

Walker ML, Gordon TP, Wilson ME (1983): Menstrual cycle characteristics of seasonally breeding rhesus monkeys. Biol Reprod 29:841–848.

Wilson ME, Gordon TP, Bernstein IS (1978): Timing of births and reproductive success in rhesus monkey social groups. J Med Primatol 7:202–212.

Wolfheim JH, Rowell TE (1972): Communication among captive talapoin monkeys (*Miopithecus talapoin*). Folia Primatol 18:224–255.

Zucker EL (1985): Inter-gender interaction patterns in captive groups of rhesus macaques. Am J Primatol 8:373–374.

Zucker EL, Kaplan JR (1981a): Allomaternal behavior in a group of free-ranging patas monkeys. Am J Primatol 1:57–64.

Zucker EL, Kaplan JR (1981b): A reinterpretation of "sexual harassment" in patas monkeys. Anim Behav 29:957–958

Comparative Behavior of African Monkeys, pages 175–195

9

The Sexual Behavior of African Monkeys and the Question of Estrus

James Loy

Department of Sociology and Anthropology, University of Rhode Island, Kingston, Rhode Island 02881

INTRODUCTION

It is not my purpose in this review to present an encyclopedic description of the sexual behavior of African monkeys. Due to numerous field and laboratory studies over the past 25 years, a large body of information on African monkey reproduction is available. Add to this the data on ecology, demography, and life histories that provide the necessary context for understanding sexual behavior, and one is faced with an emormous quantity of material—much more than can be reviewed in a short chapter. Fortunately, I was preceded in the task of reviewing the African material by several able colleagues. Reproductive cycles in general were reviewed by Hafez [1971] and Butler [1974], and an overview of copulatory behavior was provided by Michael and Zumpe [1971]. Rowell [1972] reviewed the sexual cycles and behavior of females, while Michael et al [1973] focused on male sexual behavior. Most recently, the mating strategies of male baboons were examined by Berenstain and Wade [1983], and correlations between dominance and reproductive success in male and female catarrhines were reviewed by Fedigan [1983]. Rather than attempt simply to update the material presented by my predecessors, I have chosen to use my allotted space to address and answer a single question: Do nonhuman catarrhine females show estrous cycles? Following the theme of this book, I will rely primarily on data from African monkeys for an answer. Supplementary comparative material will be taken from studies of the sexual behavior of Asian monkeys and of apes.

THE QUESTION OF ESTRUS

Since I can well imagine some readers being puzzled by the subject of this essay (after all, monkey and ape sexual behavior has been described in terms of estrous cycling for over 50 years), I will begin with an explanation for my choice of topic. As will become apparent below, I am going to argue that Old World monkeys and apes do not show estrous cycles and therefore their sexual behavior should no longer be described using estrous terminology. I have reached this conclusion after personal study of two catarrhine species (*Macaca mulatta* and *Erythrocebus patas*) and extensive reading of other scientists' works on primate sexual behavior. I have come to believe that primatologists are unconsciously sacrificing accuracy of description and communication in order to continue using estrous terminology. Due to the fact that the catarrhines' sexual behavior is usually not limited to true estrus, we have coined a confusing array of special qualifying terms [e.g., perimenstrual estrus (Loy, 1970), postconception estrus (Conaway and Koford, 1965; Kaufmann, 1965), pseudoestrus (Hrdy, 1977)] in order to fit their behavior into the estrous model. The sheer multitude of these terms suggests that we are no longer dealing with estrous cycling species, but with a different type of sexuality.

To take the problem one step further, inaccuracies and confusion among primatologists can have wide effects, especially since theories about human evolution are often based on the behavior of living primates. When primatologists describe monkey and ape sex in terms of estrous cycles and human sex in terms of continuous activity across the menstrual cycle, the gap between nonhumans and humans is made to seem enormous. In reality, human sexuality differs from monkey and ape sexuality in degree rather than in kind [evidence given below; see also Spuhler, 1979]. This closer relationship will be greatly clarified once the use of estrous terminology has been abandoned with regard to all catarrhines.

Having explained my interest in the question of estrus, I will now proceed to describe the variety of catarrhine reproductive cycles. Many of these cycles are physiological (e.g., cycles of the gonadotropins or ovarian steroids), others are histological (e.g., the menstrual cycle), and still others are morphological (e.g., sex skin color or perineal swelling cycles). The estrous cycle belongs to a fourth category, that of behavioral fluctuations. In order to evaluate the concept of estrus as applied to catarrhine primates, one must have an understanding of the first three cycle types. Therefore, the next section contains descriptions of the menstrual cycle and its accompanying hormonal fluctuations, and the morphological and behavioral effects of the ovarian hormones. General data on hormonal cycles were drawn from Wolf and Meyer [1971], Hafez [1971], King and Westfahl [1978], and Chalmers [1980]. Useful diagrams of the menstrual cycle and its accompanying hormones are presented by Hafez [1971] and Chalmers [1980].

HORMONAL AND MORPHOLOGICAL CYCLES

The menstrual cycles of female African monkeys average just over 30 days in length [Hafez, 1971; Butler, 1974]. Following Wolf and Meyer [1971], the menstrual cycle can be divided into four phases: follicular phase, ovulation, luteal phase, and menstruation. During the follicular phase, approximately the first third of the cycle, follicle stimulating hormone (FSH) from the anterior pituitary gland stimulates follicular growth within the ovary. As the follicle grows, it produces ever increasing amounts of the estrogens and small amounts of progesterone. Under the influence of the ovarian hormones, the uterine endometrium begins to thicken from the thin basal layer remaining after the previous menstruation. Luteinizing hormone (LH), another product of the pituitary gland, is present at low levels throughout most of the follicular phase, as is testosterone from the ovary.

Ovulation occurs near the middle of the menstrual cycle and is typically accompanied by sharp, but transitory, peaks in the estrogens and LH. Progesterone and testosterone may also show small peaks at this time.

Endometrial growth continues into the luteal phase under the influence of increasing levels of progesterone. This hormone, and others of ovarian origin, are produced primarily by the corpus luteum. The functional life of the corpus luteum is fairly regular and somewhat longer than the follicular phase of the menstrual cycle. The luteal phase, therefore, may account for one half or more of the total cycle length. Estrogens decline after the ovulatory peak and are present at varying levels throughout the luteal phase. LH levels during the luteal phase are generally somewhat higher than those of the follicular phase, although much lower than the peak levels reached at ovulation. In contrast, testosterone levels may be lower during the luteal phase than prior to ovulation.

The shedding of the uterine lining (menstruation) is correlated with declining levels of progesterone and usually with low levels of the estrogens and testosterone. Length of menstruation varies between species and among the females within a species, but is usually only a matter of a few days.

The morphological fluctuations shown by some African monkey species are closely correlated with the menstrual cycle and its various hormones (Table I). The most common form of morphological cycle is swelling of the

TABLE I. Morphological Reproductive Cycles of Female African Monkeys

Species	Type of cycle	Length of cycle	Length of period of maximum swelling or color	Source
Macaca sylvana	Perineal swelling	—	—	Taub, 1980
Colobus badius tephrosceles	Perineal swelling	—	—	Struhsaker, 1975
Cercocebus albigena	Perineal swelling	Median and mode = 30 days	1–7 days	Rowell and Chalmers, 1970; Wallis, 1983
Cercocebus albigena	Color of swelling	Median and mode = 30 days	Ca. 7 days	Rowell and Chalmers, 1970
Cercocebus atys	Perineal swelling	Median = 34.5 days	1–2 days	Hadidian and Bernstein, 1979; Aidara et al, 1981
Miopithecus talapoin	Perineal swelling	Mean ca. 33 days	2–6 days	Scruton and Herbert, 1970
Mandrillus leucophaeus	Perineal swelling	Median = 35 days	—	Hadidian and Bernstein, 1979
Mandrillus sphinx	Perineal swelling	Median = 33.5 days	—	Hadidian and Bernstein, 1979
Papio cynocephalus	Perineal swelling	Mean = 32.5 days	—	Hausfater, 1975
Papio ursinus	Perineal swelling	Mean = 35 days	Mean = 8.6 days	Howard-Tripp and Bielert, 1978
Papio anubis	Perineal swelling	Mean ca. 35 days	—	Butler, 1974
Papio hamadryas	Perineal swelling	Mean ca. 30 days; mode = 33 days	Ca. 2–7 days	Kummer, 1968; Asdell, 1964
Theropithecus gelada	Vesicular size	24–34 days	Ca. 10 days	Alvarez, 1973; Dunbar and Dunbar, 1974; Dunbar, 1978
Theropithecus gelada	Protrusion of vulva	24–34 days	7–10 days	Alvarez, 1973; Dunbar and Dunbar, 1974; Dunbar, 1978
Theropithecus gelada	Vesicular color	24–34 days	Ca. 10 days	Alvarez, 1973; Dunbar and Dunbar, 1974; Dunbar, 1978
Allenopithecus nigroviridis	Perineal swelling	—	—	Rowell, 1972

perineum. Females of all true baboon species (*Papio* sp.) show monthly swelling cycles, with inflation and maximum turgescence coinciding with the follicular phase and the periovulatory period, respectively. Deturgescence of the sex skin is generally thought to begin about 3 days after ovulation, with the ''deturgescent phase'' corresponding to the luteal phase of the menstrual cycle [Hausfater, 1975]. The estrogens are primarily responsible for sex skin turgescence. This is shown by the timing of the swelling within the menstrual cycle and by experimental production of swelling in ovariectomized females given replacement estradiol treatment [Saayman, 1973; Bierlert and van der Walt, 1982]. Deturgescence can be experimentally stimulated by progesterone-like substances [Saayman, 1973], and presumably is triggered by luteal phase progesterone in intact animals. Swelling cycles similar to those of baboons have been reported for the drills and mandrills (*Mandrillus* sp.), talapoin monkeys (*Miopithecus talapoin*), several species of mangabeys (*Cercocebus* sp.), red colobus monkeys (*Colobus badius tephrosceles*), Allen's swamp monkey (*Allenopithecus nigroviridis*), and the lone remaining African macaque, *Macaca sylvana* (sources are given in Table I). Sexual swellings do not occur in the genera *Cercopithecus* [Rowell, 1972], *Erythrocebus* [Rowell, 1972; Loy, personal observation], or *Theropithecus* [Dunbar and Dunbar, 1974].

Theropithecus gelada, the gelada baboon, has a form of morphological sexual signaling unlike those found in any other nonhuman primate. Gelada females show cyclical changes in the development and color of cutaneous vesicles that border areas of bare skin on the perineum and chest. Maximum vesicular size and coloration occur in the middle of the menstrual cycle. Prior to menstruation, the vesicles change from pink to white and undergo deturgescence. The hypothesis that the vesicles provide information about a female's sexual state is supported by observations of males examining the beads prior to mounting [Crook, 1972; Alvarez, 1973; Dunbar and Dunbar, 1974]. Gelada females also show cycles of protrusion of the vulva (a phenomenon that differs from true sexual swelling), with full protrusion occurring at midcycle [Dunbar and Dunbar, 1974]. Duplication of visual signals of sexual state on the chest and perineum of gelada females has been suggested to be related to the seated posture used by these animals during foraging [Crook, 1972].

Cyclical changes in the coloration of the perineal skin occur in some species of African monkeys, but on the whole such cycles are considered to be less important than sexual swelling. Female grey-cheeked mangabeys (*Cercocebus albigena*) show changes in the color of their sexual swellings from grey during inflation to bright pink at full swelling, and finally to grey again when fully deflated [Rowell and Chalmers, 1970]. Similarly, chacma baboon (*Papio ursinus*) females have been reported to show an increase in perineal pinkness coincident with full swelling [Saayman, 1970]. Color cycles, like swelling cycles, are not found in the genera *Cercopithecus* and *Erythrocebus* [Rowell, 1972].

ESTRUS DEFINED

Having reviewed the data on menstrual, hormonal, and morphological reproductive cycles, the evidence for estrous cycling among the African monkeys can now be considered. As defined by Heape in 1900, estrus "is the special period of sexual desire of the female; it is during estrus, and only at that time, the female is willing to receive the male and fruitful coition rendered possible in most, if not all mammals" [as quoted by Butler, 1974]. The fact that Heape's definition is still commonly accepted by ethologists is shown by a recent description of estrus: "When not in estrus ('in heat'), a female repels a male's advances; when she is in estrus, she invites them. Behavioral estrus occurs at the time of ovulation: thus a female allows insemination only when it is likely to be effective" [Barnett, 1981, p. 280]. Based on these definitions, estrus has the following features: 1) copulation occurs only at this time—when a female is ancstrous, she is sexually inactive; 2) during estrus the female is *willing* to copulate—she need not be coerced by the male; and 3) copulation at this time can result in conception, i.e., estrus normally coincides with ovulation [see also Fox and Laird, 1970]. A fourth feature might also be added from empirical evidence: 4) estrus is generally restricted to a brief time period [Fox and Laird, 1970; Daly and Wilson, 1978].

Circumscribed periods of sexual behavior fitting the classical definition of estrus are common among the nonprimate mammals. Norway rat (*Rattus norvegicus*) females show an estrous cycle of 4 to 6 days, with heat lasting between 12 and 20 hours [Asdell, 1964; Fox and Laird, 1970]. If a female is not in estrus, she will not allow intromission. She may kick the male off if he tries to mount, or she may simply move away from the male [Barnett, 1975]. Guinea pigs (*Cavia parellus*) have an estrous cycle of about 16 days, and the female is sexually active for about 8 hours around ovulation [Phoenix 1970; Fox and Laird, 1970]. Among domestic cattle (*Bos taurus*), the estrous cycle is usually about 20–21 days in length, and estrus itself lasts less than 1 day [Asdell, 1964; Hafez et al, 1969]. Female dogs (*Canis familiaris*) come into estrus at intervals of 4 to 18 months, and mate for periods of 4 to 14 days [Kirk, 1970].

Within the order Primates, true estrous cycles are characteristic of most, if not all, prosimians [see reviews by Butler, 1974; Van Horn and Eaton, 1979]. For example, galago (*Galago senegalensis*) females come into estrus about every 30 to 40 days and copulate for a period of 1 to 3 days. Similarly, dwarf lemurs (*Cheirogaleus major*) copulate for 2 to 3 days at approximately 30-day intervals. Ringtailed lemurs (*Lemur catta*) show an interestrus interval of about 40 days and an average duration of estrus of 21 to 22 hr [Van Horn and Eaton, 1979]. Like some nonprimate mammals [e.g., the chinchilla, *Chinchilla* sp., Weir, 1970], several species of prosimians, including galagos and the lemurs *C. major* and *Microcebus murinus,* experience opening of the vaginal orifice only at estrus and parturition. At other times, the mouth of the vagina is imperforate or nearly so [Butler, 1974; Van Horn and Eaton, 1979].

These descriptions are in close agreement with the definitions given earlier and show that true estrous cycling as exhibited by many nonprimate mammals and prosimian primates is characterized by relatively long periods of sexual quiescence (anestrus) interspersed with relatively brief periods during which copulation occurs (estrus). Sexual activity coincides with ovulation and rarely, if ever, occurs when the female is incapable of conceiving.

AFRICAN MONKEY SEXUAL BEHAVIOR

Among the African monkeys, those species that show monthly swelling cycles provide perhaps the best evidence for estrus in the original sense [Rowell, 1972]. Within a captive group of mangabeys (*Cercocebus albigena*), copulations (i.e., mounts with intravaginal ejaculation) were observed to occur only when females were at or near full swelling [Rowell and Chalmers, 1970; Chalmers and Rowell, 1971]. Copulations ceased abruptly after the onset of deturgescence of the sex skin [Rowell, 1972]. The same general pattern was noted for wild mangabeys (*C. albigena johnstonii*) by Wallis [1983]. The wild females showed a swelling cycle of about 30 days, with 1 to 3 days of full swelling at midcycle. Females presented to males most frequently on the day(s) of peak swelling, and 58% of all intromitted mounts coincided with full swelling. All intromitted mounts occurred between 2 days before and 7 days after peak swelling [Wallis, 1983]. The implication of these reports is that mangabey females mate only around the time of ovulation. Questions are raised about the reliability of this conclusion, however, by Hadidian and Bernstein's [1979] report of postconception swelling and sexual activity in captive *Cercocebus atys*.

Among talapoin monkeys, inflation of the sex skin takes about 11 days [Herbert, 1970] and maximal swelling lasts 2 to 6 days [Scruton and Herbert, 1970]. Males copulate mostly with females who are fully swollen [Scruton and Herbert, 1970; Rowell and Dixson, 1975], although males mount or half-mount females frequently during the period of inflation [i.e., the early follicular phase, Herbert, 1970; Scruton and Herbert, 1970]. In captivity, mounting is rare during the luteal phase of the menstrual cycle [Scruton and Herbert, 1970]. Within wild groups, adult males occasionally copulate with females who have small swellings, detumescing swelling, or no swelling at all. Further, there are reports of nulliparous, captive talpoin females copulating in early pregnancy [Rowell and Dixson, 1975].

Both species of *Mandrillus* (*M. leucophaeus,* the drill, and *M. sphinx,* the mandrill) show swelling cycles, and there are reports for both species of swellings accompanied by sexual activity during early pregnancy [Hadidian and Bernstein, 1979]. In addition, Struhsaker [1969] observed ejaculate on the hindquarters of a wild adult female drill who was not swollen.

By far the best studied group of African monkeys characterized by periovulatory swelling is the genus *Papio,* the baboons. These animals have also been described as mating "almost exclusively during a midcycle estrus" [Rowell, 1972, p. 82], but a review of the data suggests that the evidence for classical estrous cycling is equivocal. When one considers only interactions between females and fully adult males, especially high-ranking males, it does appear as if savannah baboons experience a brief, periovulatory period of sexual activity [*Papio anubis:* Rowell, 1967, 1970a; Hall and DeVore, 1965; *Papio ursinus:* Hall, 1962; Saayman, 1970; Hall and DeVore, 1965; Seyfarth, 1978; *Papio cynocephalus:* Hausfater, 1975]. Note, however, that here estrus is being defined in terms of sexual interactions between a female and a particular class of males (adult, dominant). This is a clear departure from the original definition, which was more broadly based on general sexual desire and willingness to mate. When interactions between females and all classes of males are considered, the picture of a clear, midcycle estrus is considerably blurred. Saayman [1970] reported that in the wild, inflating chacma (*P. ursinus*) females were mounted about as often as fully swollen females when all male partners were included. Juvenile and subadult males accounted for almost all of the mounts of inflating and deflating females, while adult males performed most mounts of fully swollen females. Similar observations have been reported for *P. anubis* [Hall and DeVore, 1965]. Saayman [1975] has estimated that a baboon female may be attractive and receptive to males for a *minimum* of the first half of her menstrual cycle (about 20 days). Furthermore, Saayman [1975] reported that chacma females may swell and copulate in early pregnancy.

Hamadryas baboons (*Papio hamadryas*), with their unimale, multifemale breeding units, may provide better evidence for classical estrus than savannah baboons. Kummer [1968] reported that copulations between a unit leader and his females occur only when the females are swollen (i.e., approximately during the second week of their 30-day swelling cycle).

Further, judging from the examples given by Kummer [1968], copulations with subadult and juvenile males from outside the unit also occur only when the female is swollen. Mounts of anestrous (i.e., deflated) females are rare, usually do not involve intromission, and may be associated with tension in the unit.

Patas monkey (*Erythrocebus patas*) females show neither sexual swelling nor sex skin color cycles. They do show a distinctive copulation solicitation, and this display plus heterosexual mounts has been used to classify females as estrous or anestrous [Loy, 1981]. Within a corral-housed patas group, it was found that periods of sexual activity ranged in length from 1 day to over 60 days, were probably socially induced in many cases, occasionally occurred near menstruation, and failed to occur in a truly cyclical fashion [Loy, 1975, 1981]. Furthermore, pregnant females in this captive group were frequently seen soliciting and copulating with the group's adult male [Loy, 1981].

Several species of *Cercopithecus* monkeys have been studied both in the wild and in captivity. Vervet (*Cercopithecus aethiops*) females have been reported to mate in all reproductive states (cycling, pregnant, lactating). Cycling females show a pattern of continuous sexual activity [Rowell, 1970b, 1971; Struhsaker, 1967]. Sykes' monkey (*Cercopithecus mitis*) females may copulate at all stages of the menstrual cycle with the exception of the late luteal phase. Mating occurs in bouts of several successive days, with mating periods separated by several days without sexual activity. These mating bouts are not, however, necessarily correlated with ovulation, although more mating occurs in the middle of the menstrual cycle than at either end. Sykes' females may continue to copulate after they have conceived [Rowell, 1970b, 1971].

Wild female redtail monkeys (*Cercopithecus ascanius*) show periods of sexual activity (i.e., heterosexual following or association, female solicitions, copulation) that average about 7 days in length, but vary from 1 to 28 days. Since the minimum recorded interval between such periods is 11 days, it appears that cycling redtail females do not limit their sexual activity to periods of fertility. Furthermore, there is some evidence for postconception sexual activity among redtails [Cords, 1984].

The sole African representative of the genus *Macaca* is the Barbary macaque (*M. sylvana*) of Morocco and Algeria. Wild Barbary females show clear cycles of perineal swelling, and peak frequencies of consort formation and copulation coincide with maximum turgescence [Taub, 1980]. In captivity, Barbary females are described as showing periods of sexual activity that are separated by periods of inactivity. Sexually active periods, however, occasionally last for 4 to 6 weeks, suggesting that they can include infertile portions of the menstrual cycle [Küster and Paul, 1984]. Sexual activity during pregnancy and lactation has also been reported for captive *M. sylvana* [Küster and Paul, 1984].

The final species to be mentioned is the gelada baboon. As noted earlier, gelada females do not show swelling cycles, but they do show cyclical fluctuations in protrusion of the vulva and in the size and color of the cutaneous vesicles that border their bare perineal and chest areas. Observations of the sexual behavior of wild geladas indicate that female solicitation calls and copulations occur at all stages of the menstrual cycle, with their highest rates at midcycle. Ejaculation rate and mounting rate both decline strongly late in the menstrual cycle [Dunbar and Dunbar, 1974; Dunbar, 1978, 1984].

The data on African monkey sexual behavior are summarized in Table II. Despite the brevity of the review, it appears that sufficient data have been presented to determine whether or not the African monkeys can be described generally as estrous cyclers. Of the 14 species included in Table II, only two (14.3%) appear to limit copulations strictly to the ovulation phase of the menstrual cycle. Even if the three least-studied species (*C. atys, M. leucophaeus,* and *M. sphinx*) are excluded from the analysis, the proportion of species whose sexual behavior fits the estrous model increases to only 18.2%. The conclusion is clear that in the vast majority of African monkey species, females do not limit their sexual behavior to estrus as it was originally defined, and as it is shown by prosimians and nonprimate mammals. Females of most species may be sexually active at many stages of the menstrual cycle, and may continue to copulate even while pregnant and/or lactating. Granted, there is considerable interspecies variation [Rowell, 1972]. Some species characterized by perineal swellings give the impression of being classic estrous cyclers. Under close examination, however, even animals such as mangabeys, talapoins, and baboons are found to engage in sexual behavior at times other than at ovulation.

Without going into detail, it should be noted that findings similar to those above have been reported for Asian monkeys [Rowell, 1972; Butler, 1974; Slob et al, 1978; Keverne, 1981; Hrdy, 1981], thus confirming the conclusion that, by and large, cercopithecoid sexual behavior does not conform to the estrous model [see also Saayman, 1975]. Since this is the case, it is clear that primatologists have little to gain and considerable to lose by continuing to describe Old World monkey sexual behavior in estrous terminology. A new model with new terms is required by the more complex sexuality of these animals.

TABLE II. Sexual Behavior of Female African Monkeys

Species	Morphological cycle	Heterosexual sexual activity observed during:[a]					
		Foll. phase	Ovulation	Luteal phase	Menst.	Pregnancy	Lactation
Cercocebus albigena	Perineal swelling	—	Cop (c,w)	—	—	—	—
Cercocebus atys	Perineal swelling	—	—	—	—	Unspec. sex. act. (c)	—
Miopithecus talapoin	Perineal swelling	Freq. mt (c); occas. mt (w)	Cop (c,w)	Rare mt (c,w)	—	Cop (c)	—
Mandrillus leucophaeus	Perineal swelling	—	—	—	—	Unspec. sex. act. (c)	—
Mandrillus sphinx	Perineal swelling	—	—	—	—	Unspec. sex. act. (c)	—
Papio ursinus	Perineal swelling	Freq. mt (w)	Freq. mt and cop (w)	Mt (w)	—	Cop (w)	—
Papio anubis	Perineal swelling	Mt (w)	Freq. mt and cop (w)	Mt (w)	—	—	—
Papio hamadryas	Perineal swelling	—	Cop (w)	—	—	—	—
Erythrocebus patas	None	Sol (c); occas. cop (c)	Sol (c); cop (c)	—	Rare cop (c)	Sol (c); cop (c)	—
Cercopithecus aethiops	None	Cop (c)	Cop (c)	Cop (c)	Cop (c)	Cop (c)	Occas. cop (c)
Cercopithecus mitis	None	Cop (c)	Cop (c)	Cop (c)	Cop (c)	Occas. cop (c)	—
Cercopithecus ascanius	None	(Long periods of sexual activity suggest some copulations during infertile phases)				Occas. cop (w)	—
Macaca sylvana	Perineal swelling	—	Cop (w)	—	—	Cop (c)	Cop (c)
Theropithecus gelada	Vesicle size and color; protrusion of vulva	Sol call and cop (w)	Sol call and cop (w)	Sol call and cop (w)	Rare sol call and cop (w)	—	—

[a]Cop: copulation; mt: male on female mount (may not include intravaginal ejaculation); sol: female solicitation to male; sol call: female solicitation call to male; unspec. sex. act.: unspecified ''sexual activity''; (c): data from captive animals; (w): data from wild animals; occas.: occasional. It is very likely that many of the blank spaces in the table are the result of missing data rather than the actual absence of sexual behavior during a particular reproductive stage.

APR MODEL

Fortunately, a most acceptable alternative model was proposed a few years ago by Beach [1976]. In this model, female sexuality is divided into three aspects or categories: attractiveness, proceptivity, and receptivity (hence the APR model). Attractiveness refers to the female's stimulus value in evoking sexual responses from males. Physical characteristics, such as perineal swelling and vaginal odor, may enhance attractiveness, but so too may aspects of behavior. Proceptivity refers to behaviors used by females to establish or maintain a sexual interaction. Patas monkey females, for example, have a characteristic solicitation display [Loy, 1981], which they use to initiate copulations. Receptivity is defined in terms of the female's cooperation in copulation. A rhesus female who willingly presents her rear to the male in response to his touch, and who then holds her stance while the male mounts and ejaculates intravaginally would be considered receptive.

All three aspects of the APR model may be influenced by both external variables (environmental and social factors) and internal variables (especially the steroid hormones). Generally speaking, among nonhuman primates all three aspects of female sexuality seem to be stimulated by the hormonal conditions around ovulation (relatively high levels of the estrogens and testosterone, and low levels of progesterone), and to be dampened by the hormonal conditions during infertile phases [Chalmers, 1980; Gordon, 1981; Baum, 1983]. These hormonal effects are most clearly seen when the sexual behavior of true estrous cyclers is described in APR terms. Among prosimians and many nonprimate mammals, attractiveness, proceptivity, and receptivity are manifest coincidentally and at only one point in the female's reproductive cycle—around ovulation. During infertile phases, all three aspects decline in intensity to near zero levels.

The sexual behavior of female cercopithecoids differs from that of estrous cyclers primarily in the relaxation of strict hormonal control of proceptivity and receptivity. Although there is considerable interspecies variation, it seems reasonable to say that proceptivity and receptivity have become strongly situation-dependent [term taken from Hrdy, 1981, who also gives a functional explanation for these behavioral changes]. Regardless of her reproductive phase or hormonal condition, given the right set of circumstances, an Old World monkey female can display proceptive behaviors and engage in copulation. She is most likely to be proceptive and receptive around ovulation, of course, since hormonal stimulation has not been lost but has simply become less essential for the expression of sexual behavior.

The sexual attractiveness of cercopithecoid females seems to be more closely tied to hormonal cycles than is the case for proceptivity and receptivity. This is obviously due to the fact that in many species female attractiveness is largely a function of perineal swelling, sex skin color, vaginal odor, or some other physical attribute. Nonetheless, in several species, some with and some without clear morphological cycles, females are sufficiently attractive to males to evoke mounts and copulations throughout much of the menstrual cycle and during pregnancy [Table II; see also the data on *Macaca arctoides:* Slob et al, 1978; Slob and Nieuwenhuijsen, 1980].

To summarize the argument thus far: most cercopithecoids do not show estrous cycles, in which all aspects of female sexuality appear only around ovulation and are obviously under strict hormonal control. Rather, hormonal control over proceptivity and receptivity (and possibly attractiveness) has been relaxed somewhat and at least the first two aspects of female sexuality have become strongly situation-dependent.

There are other possible interpretations of the data on cercopithecoid sexuality, of course. Keverne [1976, 1981] and his co-workers have argued that Old World monkey females are continuously receptive (and, by inference, continuously attractive) throughout the menstrual cycle. This theory views receptivity as determined by constant adrenal androgens rather than cyclic ovarian hormones and is based largely on laboratory results showing that throughout their menstrual cycles female rhesus macaques will bar-press to gain access to a male, and that once paired with a male, females will usually cooperate with his attempts to copulate.

I am skeptical about the theory of continuous receptivity for several reasons. First, I do not believe that either female apes or women are continuously receptive (see below), and thus it would seem unlikely that cercopithecoid females would show this trait. Second, there surely are many reasons why a social animal like a rhesus female would work to gain access to a male conspecific—sexual desire is only one. To date, laboratory workers have failed to show that their findings are not influenced by nonsexual sociality. Third, much of the evidence for continuous receptivity comes from monkeys pair-tested in small cages. In this situation, the sexual communication of the monkeys may be impaired and the behavioral options available to the female may be limited [Wallen, 1982]. Laboratory workers may have mistaken an extreme form of situation-dependent receptivity for hormonally-based continuous receptivity. It is difficult to determine whether the sexual cooperation shown by a macaque female during a pair-test is related to her hormonal state or the fact that she is faced with a large, insistent male and no means of escape. Here she may be willing to copulate (and thus is technically receptive) only because she has no choice but compliance [Wallen and Goy,

1977; Goldfoot, 1977; Wallen, 1982]. I suggest that until we can reliably separate hormonal effects from test condition effects, we should interpret pair-test results using the most conservative of the alternative theories, i.e., the theory of situation-dependent receptivity.

If one accepts the conclusion that cercopithecoids are not estrous cyclers, but rather show a reduction in hormonal influence on the APR variables and the development of situation-dependent proceptivity and receptivity, then the question becomes: What about apes and humans? At the beginning of the essay, I suggested that monkey, ape, and human sexualities differ in degree rather than in kind. A very brief review of the data on hominoids will show why I think this is true.

HOMINOID SEXUAL BEHAVIOR

Only the great apes will be included in this review. The sexual behavior of the Hylobatidae (gibbons and siamangs) is poorly known, and they are rather distant evolutionary relatives of humans [Oxnard, 1984]. Orang-utans (*Pongo pygmaeus*) are probably also distant relatives of humans [Oxnard, 1984], but their sexual behavior will be described since it is often compared to that of human beings [e.g., Short, 1979]. In the wild, copulations between cycling orang females and adult males tend to occur during short consortships that probably coincide with ovulation. Female proceptivity also appears during these consortships [Galdikas, 1981]. Males, especially subadults, may, however, force copulation at any stage of the menstrual cycle or during pregnancy [Galdikas, 1981; Mitani, 1985]. The cyclic interest of females in sex is clearly shown by laboratory pair-tests using a caging arrangement that allows escape from the male. In this situation, copulations occur mainly around ovulation [Nadler, 1982]. We may conclude that among orangs, female attractiveness is dependent on a male's age, and subadult males may find females constantly attractive; proceptivity is under strong hormonal control and therefore tied to ovulation; and receptivity is under moderate hormonal influence and to some extent situation-dependent.

Observations of gorilla (*Gorilla gorilla*) females living in heterosexual groups suggest that attractiveness, proceptivity, and receptivity peak coincidentally at ovulation, and that copulations do not occur during other phases of the menstrual cycle [Hess, 1973; Harcourt and Stewart, 1978; Dixson, 1981; Fossey, 1982; Mitchell et al, 1985]. Pair-test data, however, indicate that occasionally males may find a female attractive throughout most of her menstrual cycle (excluding only the late luteal phase), and that receptivity is to some degree situation-dependent [Nadler et al, 1983]. Copulations during pregnancy have also been reported both from wild animals [Harcourt and Stewart, 1978] and from captives [Hess, 1973].

Common chimpanzee (*Pan troglodytes*) females show large perineal swellings, and all observers agree that the sexual behavior of a cycling female peaks when she is maximally swollen [i.e., around ovulation, McGinnis, 1973, as cited by Short, 1979; Savage-Rumbaugh, and Wilkerson, 1978; Coe et al, 1979; Tutin, 1979; Wallis, 1982]. Copulations may occur, however, during nonfertile phases of the menstrual cycle (when the female shows little, if any, swelling) and during pregnancy [Coe et al, 1979; Tutin, 1979; Wallis, 1982]. In general, it appears that chimpanzee attractiveness and proceptivity may be under moderate hormonal control, but that receptivity is strongly situation-dependent and may be shown during virtually any reproductive stage [Wallis, 1982].

Much less is known about the pygmy chimpanzee (*Pan paniscus*) than *P. troglodytes,* but the few published accounts suggest that pygmy chimp females are sexually active throughout the menstrual cycle [Savage-Rumbaugh and Wilkerson, 1978; Kano, 1982]. In particular, attractiveness and receptivity may be nearly constant characteristics of female pygmy chimps [Kano, 1982].

To summarize the data on great ape sexual behavior: attractiveness and proceptivity are under at least moderate hormonal influence and therefore tend to occur (or peak) at ovulation. Under some circumstances, however, some males may find females always attractive. Receptivity may be under somewhat less hormonal influence than the other aspects of female sexuality and should be regarded as moderately-to-strongly situation-dependent.

Reliable data on human sexual behavior are notoriously difficult to obtain, but virtually all authors agree that women are constantly attractive to men, and that they can behave proceptively and display receptivity at any time during the menstrual cycle, during pregnancy, and during lactation [see summaries in Beach, 1978; Symons, 1979; Hrdy, 1981]. Constant attractiveness is related at least partially to the presence of permanent (i.e., noncyclic) physical stimuli such as prominent breasts [Beach, 1978; note, however, that authorities disagree about the theory that the human breast evolved to provide sexual stimulation for males—compare Symons, 1979, with Short, 1979]. Proceptivity and receptivity are probably influenced to only a small extent by gonadal hormones (although perhaps more so than is the case for attractiveness), but neither characteristic is constantly displayed by women [Beach, 1978]. With particular regard to receptivity, Beach [1978] has noted that while women are always copulable and sexually arousable, they are not constantly willing to engage in intercourse, and thus

TABLE III. Degrees of Hormonal Influence on the Attractiveness, Proceptivity, and Receptivity of Primates, and the Development of Situation-Dependent Proceptivity and Receptivity

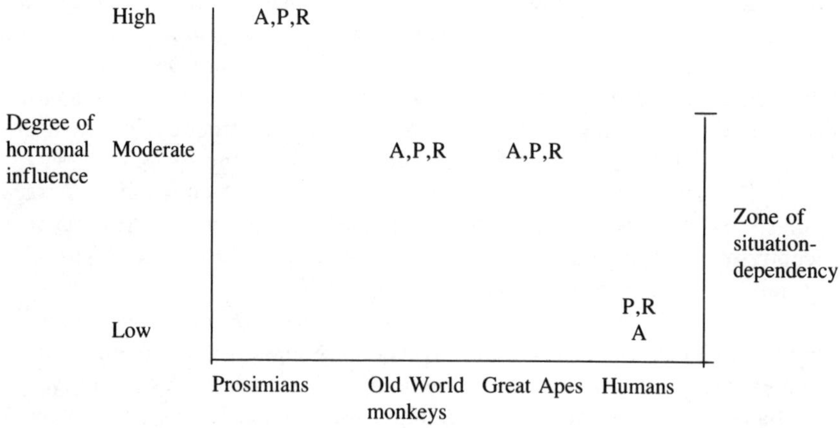

cannot be said to be continuously receptive. (It is difficult for observers of monkeys and apes to recognize and measure females' willingness to copulate, and therefore this element has been dropped from primatologists' definitions of receptivity. Willingness to mate was, however, an important part of the original definition of estrus and since it can be measured among humans, it should be incorporated into definitions of women's receptivity.)

From the material on monkey, ape, and human sexual behavior reviewed above, the major changes that occurred during the evolution of female sexuality among the catarrhines can be outlined (see Table III). The primary change from the prosimians to cercopithecoids and great apes was a shift from strict hormonal control of attractiveness, proceptivity, and receptivity to moderate hormonal control of all three factors plus the appearance of situation-dependent proceptivity and receptivity. The ape-to-human shift involved a further strong reduction in hormonal influence on all aspects of female sexuality. Attractiveness became continuous, and proceptivity and receptivity became even more obviously situation-dependent. Clearly, this view of catarrhine sexual evolution adheres more closely to the known facts and should be more useful in stimulating further investigation than the older and erroneous view that monkeys and apes still show estrus and only humans have lost it.

CONCLUSIONS

1. Most, if not all, species of Old World monkeys and apes are inaccurately described as showing estrous cycles. In addition to mating around ovulation, cercopithecoid and ape females may copulate during infertile phases of the menstrual cycle, during pregnancy and during lactation.

2. Since Old World monkeys and apes are not estrous cyclers, primatologists should stop describing them as such. If we continue to apply estrous terminology to these animals we risk fostering an erroneous and simplistic view of their sexual behavior and producing inaccurate scientific reports that confuse specialists and nonspecialists alike.

3. The APR model [Beach, 1976] provides a much more useful way of viewing catarrhine female sexuality than does the estrous model. Compared to true estrous cyclers such as most prosimians (who are attractive, proceptive, and receptive simultaneously and only around ovulation), cercopithecoid and ape females have experienced a significant reduction in hormonal influence on all three APR variables. Furthermore, among Old World monkeys and apes, proceptivity and receptivity have become moderately-to-strongly situation-dependent.

3a. There is insufficient evidence to accept the theory that cercopithecoid and ape females are continuously receptive throughout the menstrual cycle.

4. Compared to monkeys and apes, women show much less hormonal influence on attractiveness, proceptivity, and receptivity. Attractiveness to males has become a constant feature of human females' sexuality. Proceptivity and receptivity, while not constantly present, are nonetheless always possibilities because they have become so thoroughly situation-dependent.

ACKNOWLEDGMENTS

I thank Professor C. B. Peters for critically reviewing an earlier version of the essay. His comments strengthened the paper significantly. I also thank Roseann Kane for typing the paper.

REFERENCES

Aidara D, Badawi M, Tahiri-Zagret C, Robyn C (1981): Changes in concentrations of serum prolactin, FSH, oestradiol and progesterone and of the sex skin during the menstrual cycle in the mangabey monkey (*Cercocebus atys lunulatus*). J Reprod Fertil 62:475–481.

Alvarez F (1973): Periodic changes in the bare skin areas of *Theropithecus gelada*. Primates 14:195–199.

Asdell SA (1964): "Patterns of Mammalian Reproduction." Ithaca: Cornell University Press.

Barnett SA (1975): ''The Rat.'' Chicago: The University of Chicago Press.
Barnett SA (1981): ''Modern Ethology.'' New York: Oxford University Press.
Baum MJ (1983): Hormonal modulation of sexuality in female primates. Bioscience 33:578–582.
Beach FA (1976): Sexual attractivity, proceptivity, and receptivity in female mammals. Horm Behav 7:105–138.
Beach FA (1978): Human sexuality and evolution. In Washburn SL, McCown, ER (eds): ''Human Evolution: Biosocial Perspectives.'' Menlo Park, CA: Benjamin/Cummings, pp 122–153.
Berenstain L, Wade TD (1983): Intrasexual selection and male mating strategies in baboons and macaques. Int J Primatol 4:201–235.
Bielert C, van der Walt LA (1982): Male chacma baboon (*Papio ursinus*) sexual arousal: Mediation by visual cues from female conspecifics. Psychoneuroendocrinology 7:31–48.
Butler H (1974): Evolutionary trends in primate sex cycles. Contrib Primatol 3:2–35.
Chalmers N (1980): ''Social Behaviour in Primates.'' Baltimore: University Park Press.
Chalmers NR, Rowell TE (1971): Behaviour and female reproductive cycles in a captive group of mangabeys. Folia Primatol 14:1–14.
Coe CL, Connolly AC, Kraemer HC, Levine S (1979): Reproductive development and behavior of captive female chimpanzees. Primates 20:571–582.
Conaway CH, Koford CB (1965): Estrous cycles and mating behavior in a free-ranging band of rhesus monkeys. J Mammalogy 45:577–588.
Cords M (1984): Mating patterns and social structure in redtail monkeys (*Cercopithecus ascanius*). Z Tierpsychol 64:313–329.
Crook JH (1972): Sexual selection, dimorphism, and social organization in the primates. In Campbell B (ed): ''Sexual Selection and the Descent of Man 1871–1971.'' Chicago: Aldine Publishing Company, pp 231–281.
Daly M, Wilson M (1978): ''Sex, Evolution, and Behavior.'' North Scituate, MA: Duxbury Press.
Dixson AF (1981): ''The Natural History of the Gorilla.'' New York: Columbia University Press.
Dunbar RIM (1978): Sexual behaviour and social relationships among gelada baboons. Anim Behav 26:167–178.
Dunbar RIM (1984): ''Reproductive Decisions.'' Princeton: Princeton University Press.
Dunbar RIM, Dunbar P (1974): The reproductive cycle of the gelada baboon. Anim Behav 22:203–210.
Fedigan LM (1983): Dominance and reproductive success in primates. Yearbook Phys Anthropol 26:91–129.
Fossey D (1982): Reproduction among free-living mountain gorillas. Am J Primatol, Suppl 1:97–104.
Fox RR, Laird CW (1970): Sexual cycles. In Hafez ESE (ed): ''Reproduction and Breeding Techniques for Laboratory Animals.'' Philadelphia: Lea and Febiger, pp 107–122.
Galdikas BMF (1981): Orangutan reproduction in the wild. In Graham CE (ed): ''Reproductive Biology of the Great Apes.'' New York: Academic Press, pp 281–300.
Goldfoot DA (1977): Sociosexual behaviors of nonhuman primates during development and maturity: Social and hormonal relationships. In Schrier A (ed): ''Behavior Primatology: Advances in Research and Theory.'' Hillsdale, NJ: Lawrence Ehrlbaum, pp 139–184.
Gordon TP (1981): Reproductive behavior in the rhesus monkey: Social and endocrine variables. Am Zool 21:185–195.

Hadidian J, Bernstein IS (1979): Female reproductive cycles and birth data from an Old World monkey colony. Primates 20:429–442.

Hafez ESE (1971): Reproductive cycles. In Hafez ESE (ed): "Comparative Reproduction of Nonhuman Primates." Springfield: Charles C. Thomas, pp 160–204.

Hafez ESE, Schein MW, Ewbank R (1969): The behaviour of cattle. In Hafez ESE (ed): "The Behavior of Domestic Animals." Baltimore: Williams and Wilkins, pp 235–295.

Hall KRL (1962): The sexual, agonistic and derived social behaviour patterns of the wild chacma baboon, *Papio ursinus*. Proc Zool Soc Lond 139:283–327.

Hall KRL, DeVore I (1965): Baboon social behavior. In DeVore I (ed): "Primate Behavior." New York: Holt, Rinehart and Winston, pp 53–110.

Harcourt AH, Stewart KJ (1978): Sexual behaviour of wild mountain gorillas. In Chivers DJ, Herbert J (eds): "Recent Advances in Primatology. Behavior." London: Academic Press, Vol. 1, pp 611–612.

Hausfater G (1975): Dominance and reproduction in baboons (*Papio cynocephalus*). Contrib Primatol 7:1–150.

Heape W (1900): The 'sexual season' of mammals, and the relation of the 'pro-oestrum' to menstruation. Q J Microsc Sci 44:1–70.

Herbert J (1970): Hormones and reproductive behaviour in rhesus and talapoin monkeys. J Reprod Fertil, Supplement 11:119–140.

Hess JP (1973): Some observations on the sexual behaviour of captive lowland gorillas, *Gorilla g. gorilla*. In Michael RP, Crook, JH (eds): "Comparative Ecology and Behaviour of Primates." London: Academic Press, pp 507–581.

Howard-Tripp ME, Bielert C (1978): Social contact influences on the menstrual cycle of the female chacma baboon (*Papio ursinus*). J S Afr Vet Assoc 49:191–192.

Hrdy SB (1977): "The Langurs of Abu." Cambridge: Harvard University Press.

Hrdy SB (1981): "The Woman That Never Evolved." Cambridge: Harvard University Press.

Kano T (1982): The social group of pygmy chimpanzees (*Pan paniscus*) of Wamba. Primates 23:171–188.

Kaufmann JH (1965): A three-year study of mating behavior in a freeranging band of rhesus monkeys. Ecology 46:500–512.

Keverne EB (1976): Sexual receptivity and attractiveness in the female rhesus monkey. Adv Study Behav 7:155–200.

Keverne EB (1981): Do Old World primates have oestrus? Malaysian Appl Biol 10:119–126.

Kirk RW (1970): Dogs. In Hafez ESE (ed): "Reproduction and Breeding Techniques for Laboratory Animals." Philadelphia: Lea and Febiger, pp 224–236.

Kling OR, Westfahl PK (1978): Steroid changes during the menstrual cycle of the baboon (*Papio cynocephalus*) and human. Biol Reprod 18:392–400.

Kummer H. (1968): "Social Organization of Hamadryas Baboons." Chicago: The University of Chicago Press.

Küster J, Paul A (1984): Female reproductive characteristics in semifreeranging Barbary macaques (*Macaca sylvanus* L. 1758). Folia Primatol 43:69–83.

Loy J (1970): Peri-menstrual sexual behavior among rhesus monkeys. Folia Primatol 13:286–297.

Loy J (1975): The copulatory behaviour of adult male patas monkeys, *Erythrocebus patas*. J Reprod Fertil 45:193–195.

Loy J (1981): The reproductive and heterosexual behaviors of adult patas monkeys in captivity. Anim Behav 29:714–726.

McGinnis PR (1973): Patterns of sexual behaviour in a community of free-living chimpanzees. PhD Dissertation, Cambridge University, Cambridge, England.

Michael RP, Wilson M, Plant TM (1973): Sexual behaviour of male primates and the role of testosterone. In Michael RP, Crook JH (eds): "Comparative Ecology and Behaviour of Primates." London: Academic Press, pp 235–313.

Michael RP, Zumpe D (1971): Patterns of reproductive behavior. In Hafez ESE (ed): "Comparative Reproduction of Nonhuman Primates." Springfield: Charles C. Thomas, pp 205–242.

Mitani JC (1985): Mating behaviour of male orangutans in the Kutai Game Reserve, Indonesia. Anim Behav 33:392–402.

Mitchell WR, Lindburg DG, Shideler SE, Presley S, Lasley BL (1985): Sexual behavior and urinary ovarian hormone concentrations during the lowland gorilla menstrual cycle. Int J Primatol 6:161–172.

Nadler R (1982): Laboratory research on sexual behavior and reproduction of gorillas and orangutans. Am J Primatol, Supplement 1:57–66.

Nadler RD, Collins DC, Miller LC, Graham CE (1983): Menstrual cycle patterns of hormones and sexual behavior in gorillas. Horm Behav 17:1–17.

Oxnard CE (1984): "The Order of Man." New Haven: Yale University Press.

Phoenix CH (1970): Guinea pigs. In Hafez ESE (ed): "Reproduction and Breeding Techniques for Laboratory Animals." Philadelphia: Lea and Febiger, pp 244–257.

Rowell TE (1967): Female reproductive cycles and the behavior of baboons and rhesus macaques. In Altmann SA (ed): "Social Communication Among Primates." Chicago: The University of Chicago Press, pp 15–32.

Rowell TE (1970a): Baboon menstrual cycles affected by social environment. J Reprod Fertil 21:133–141.

Rowell TE (1970b): Reproductive cycles of two *Cercopithecus* monkeys. J Reprod Fertil 22:321–338.

Rowell TE (1971): Organization of caged groups of *Cercopithecus* monkeys. Anim Behav 19:625–645.

Rowell TE (1972): Female reproduction cycles and social behavior in primates. Adv Study Behav 4:69–105.

Rowell TE, Chalmers NR (1970): Reproductive cycles of the mangabey *Cercocebus albigena*. Folia Primatol 12:264–272.

Rowell TE, Dixson AF (1975): Changes in social organization during the breeding season of wild talapoin monkeys. J Reprod Fertil 43:419–434.

Saayman GS (1970): The menstrual cycle and sexual behaviour in a troop of free ranging chacma baboons (*Papio ursinus*). Folia Primatol 12:81–110.

Saayman, GS (1973): Effects of ovarian hormones on the sexual skin and behavior of ovariectomized baboons (*Papio ursinus*) under free-ranging conditions. In Phoenix CH (ed): "Primate Reproductive Behavior." Basel: S. Karger, pp 64–98.

Saayman GS (1975): The influence of hormonal and ecological factors upon sexual behavior and social organization in Old World primates. In Tuttle RH (ed): "Socioecology and Psychology of Primates." The Hague: Mouton, pp 181–204.

Savage-Rumbaugh ES, Wilkerson BJ (1978): Socio-sexual behavior in *Pan paniscus* and *Pan troglodytes:* A comparative study. J Hum Evol 7:327–344.

Scruton DM, Herbert J (1970): The menstrual cycle and its effect on behaviour in the talapoin monkey (*Miopithecus talapoin*). J Zool (Lond) 162:419–436.

Seyfarth RM (1978): Social relationships among adult male and female baboons. I. Behaviour during sexual consortship. Behaviour 64:204–226.

Short RV (1979): Sexual selection and its component parts, somatic and genital selection, as illustrated by man and the great apes. Adv Study Behav 9:131–158.

Slob AK, Nieuwenhuijsen K (1980): Heterosexual interactions of pairs of laboratory-housed stumptail macaques (*Macaca arctoides*) under continuous observation with closed-circuit video recording. Int J Primatol 1:63–80.

Slob AK, Wiegand SJ, Goy RW, Robinson JA (1978): Heterosexual interactions in laboratory-housed stumptail macaques (*Macaca arctoides*): Observations during the menstrual cycle and after ovariectomy. Horm Behav 10:193–211.

Spuhler JN (1979): Continuities and discontinuities in anthropoid-hominid behavioral evolution: Bipedal locomotion and sexual receptivity. In Chagnon NA, Irons W (eds): "Evolutionary Biology and Human Social Behavior: An Anthropological Perspective." North Scituate, MA: Duxbury Press, pp 454–461.

Struhsaker TT (1967): Behavior of vervet monkeys (*Cercopithecus aethiops*). Univ Calif Publ Zool 82:1–74.

Struhsaker TT (1969): Correlates of ecology and social orgnization among African cercopithecines. Folia Primatol 11:80–118.

Struhsaker TT (1975): "The Red Colobus Monkey." Chicago: The University of Chicago Press.

Symons D (1979): "The Evolution of Human Sexuality." Oxford: Oxford University Press.

Taub DM (1980): Female choice and mating strategies among wild Barbary macaques (*Macaca sylvanus* L.). In Lindburg DG (ed): "The Macaques: Studies in Ecology, Behavior and Evolution." New York: Van Nostrand Reinhold, pp 287–344.

Tutin CEG (1979): Mating patterns and reproductive strategies in a community of wild chimpanzees (*Pan troglodytes schweinfurthii*). Behav Ecol Sociobiol 6:29–38.

Van Horn RN, Eaton GG (1979): Reproductive physiology and behavior in prosimians. In Doyle GA, Martin RD (eds): "The Study of Prosimian Behavior." New York: Academic Press, pp 79–122.

Wallen K (1982): Influence of female hormonal state on rhesus sexual behavior varies with space for social interaction. Science 217:375–377.

Wallen K, Goy RW (1977): Effects of estradiol benzoate, estrone, and propionates of testosterone or dihydrotestosterone on sexual and related behaviors of ovariectomized rhesus monkeys. Horm Behav 9:228–248.

Walli J (1982): Sexual behavior of captive chimpanzees (*Pan troglodytes*): Pregnant versus cycling females. Am J Primatol 3:77–88.

Wallis SJ (1983): Sexual behavior and reproduction of *Cercocebus albigena johnstonii* in Kibale Forest, Western Uganda. Int J Primatol 4:153–166.

Weir BJ (1970): Chinchilla. In Hafez ESE (ed): "Reproduction and Breeding Techniques for Laboratory Animals." Philadelphia: Lea and Febiger, pp 209–223.

Wolf RC, Meyer RK (1971): Gonadotropins and gonadal hormones. In Hafez ESE (ed): "Comparative Reproduction of Nonhuman Primates." Springfield: Charles C. Thomas, pp 39–57.

Comparative Behavior of African Monkeys, pages 197–234

10

Management and Husbandry of African Monkeys in Captivity

Mollie Bloomstrand and Terry L. Maple

University of Texas System Cancer Center, Bastrop, Texas 78602 (M.B.), and School of Psychology, Georgia Institute of Technology, Atlanta, Georgia 30332 (T.L.M.)

INTRODUCTION

The biomedical value of nonhuman primates has been reliably demonstrated in the laboratory. As living creatures uniquely adapted to their natural habitats they are equally interesting to science. In the wild they can also be observed by tourists and hopefully protected. As we endeavor to preserve natural habitats for primates and other forms of wildlife, we should also maintain and propogate sufficient populations in captive settings. The most responsible management strategy provides for applied scientific programs in which naturalistic field observations and captive manipulations are coordinated, catalogued, and compared. This approach was long ago advocated by the comparative psychologist John Broadus Watson who wrote:

> . . . there is no conflict between field work and laboratory work. The field is both the source of problems and the place where the laboratory solutions of these problems are tested. (p. 31)

There is growing interest in an applied management science devoted to ''nonhuman factors.'' Elsewhere [Maple and Finlay, 1986] we have referred to such tasks as the work of ''applied primatology.'' The scope of this field is broad indeed. The applied primatologist is by training and inclination a problem-solver who employs the tools and collective knowledge of primatology. For example, a fieldworker may be concerned with how to protect a population of rare monkeys. Primatological data are therefore applied to define the effective range for a national park. In the laboratory and the zoo, applied primatologists are concerned with the maintenance, exhibition, husbandry, and propagation of nonhuman primates. Success in these areas requires a baseline knowledge of primates in the wild.

For the past decade, our laboratory for applied primatology has focused on the *Pongidae*. However, since we cut our professional teeth on monkeys, we have returned to this group in the present chapter. Herein we will apply our knowledge of the cercopithecines in an attempt to delineate potential and known management, exhibition and husbandry issues. We have organized data from field, lab, and zoo research. Where appropriate we will apply some generalizations derived from studies of other monkeys, apes, and even human subjects. This effort is directed to an audience of practitioners in laboratories, primate centers, and zoological parks.

THE VARIABLES OF CAPTIVITY

Captive environments are composed of many changing elements. These variables include the following: 1) spatial variables—area, volume, distance; 2) social variables—number of companions, age/sex distribution, compatibility, familiarity; 3) temporal variables—light/dark cycle, ambient temperature, humidity, feeding schedule, visual access to keepers and scarce resources; 4) environmental complexity—alien taxa, browse, cover, furniture, moveable objects, stationary objects, substrate characteristics; 5) enrichment—opportunities to work, presence of manipulanda, vegetation, water; 6) health variables—cleanliness, comfort, variety of food, ventilation, sunlight; 7) environmental stressors—intraspecific aggression, human intrusions and contact, noise, social density.

We will consider many of these variables in relation to what is known about the behavior and adaptations of cercopithecines in the wild and in captivity. The subheadings utilized herein provide selected information about *Cercopithecus aethiops* (and closely related subspecies), *Cercopithecus mitis, Cercopithecus neglectus, Erythrocebus patas,* and *Miopithecus talapoin*. In the sections to follow we will review those field and laboratory data that can provide some guidelines for managing the animals in captivity. Hopefully, this information will contribute to better reproduction, improved husbandry, reductions in mortality and morbidity, and otherwise enhance the well-being of these primates in captivity.

GENERAL TAXONOMIC CHARACTERISTICS

Cercopithecus aethiops

The vervet (*C. aethiops*) and the northwestern form (*C. aethiops sabaeus*) known as the African green monkey or grivet is a relatively small Old World

TABLE I. Distribution of Various Forms of *C. aethiops* in American Zoos[a]

C. aethiops (n = 14)	
2.5	Burnet Park Zoo, Syracuse, NY
0.2	Ellen Trout Zoo, Lufkin, TX
2.2.1	San Diego Wild Animal Park
C. a. aethiops (n = 30)	
2.4.1	Caldwell Children's Zoo, Tyler, TX
2.0	Ellen Trout Zoo, Lufkin, TX
4.1	Erie Zoological Park, Erie, PA
1.4.2	Ft. Wayne Children's Zoo, Ft. Wayne, IN
0.1	Lion Country Safari, W. Palm Beach, FL
3.5	S. Nevada Zoological Park, Las Vegas, NV
C. a. hilgerti (n = 3)	
1.1.1	Rock Island Co. Zoological Preserve, Moline, IL
C. a. pygerythrus (n = 1)	
1.0	Zoo Atlanta, Atlanta, GA
C. a. sabeus (n = 11)	
1.2	Cohanzick Zoo, Bridgeton, NJ
2.2	Lion Country Safari, W. Palm Beach, FLa
1.2	Micke Grove Zoo, Lodi, CA
1.0	Knowland Park Zoo, Oakland, CA

[a]According to the records of ISIS (as of June 30, 1986).

primate with adult males weighing an average of 7 kg and adult females averaging 5.6 kg respectively [Fedigan, 1982]. Their small size is one reason that they are regarded as promising subjects for laboratory use. They are easier to manage and less expensive to feed than most Old World forms, yet they are large enough for most biomedical studies, including reproduction, embryology, and teratology [Hess et al, 1979] (Table I).

The vervet group is extremely adaptable, residing in a wide range of habitats—savannahs, areas between woodlands and savannahs, riverine forests, rain forests, and urban parks [Hall and Gartlan, 1965; Kavanaugh, 1980]. There are reports of vervets spending up to 70% of their time in trees [Poirier, 1972], but in most areas of their range they are considered to be semiterrestrial. They often sleep in forested areas at night, sometimes spreading out over 686 m, according to Struhsaker [1967], while ranging into open areas to feed and drink during the day [Fedigan, 1982]. Daily ranges of 500–1000 m have been reliably estimated [Dunbar, 1974; Hall and Gartlan, 1965], and there is one impressive report of a group that reportedly moved 17.7 km in less than 24 hr [Brain, cited in Hall and Gartlan, 1965]. Home ranges are considerably larger with estimates of 0.184 to 0.958 km^2 in Amboseli vervets [Struhsaker, 1967].

Vervets are highly territorial in most areas of their range and members of both sexes defend their group's territory [Fedigan, 1982]. Generally a disorganized retreat occurs in response to an alarm vocalization, although occasionally they may "mob" an intruder. Captive vervets have also been known to defend a group member against caretakers [Fedigan, 1982]. General activity levels have not been measured, but Wrangham [1981] stated that one group spent an average of 26.6% of their time resting and 50.5% feeding. They seem to follow a typical primate pattern composed of: 1) slow feeding and sunning early in the morning, with a higher energy level expended by late morning; 2) little movement during the hottest, mid-portion of the day; 3) intense feeding later in the afternoon; and 4) a return to sleeping trees before dark [Poirier, 1972].

Cercopithecus mitis spp.

Only a few studies of the blue monkey (*C. mitis stuhlmanni*) have been published. One such study in the Kibale Forest of Uganda was carried out by Rudran [1978]. These animals are extremely arboreal and rarely descend to the ground. Since their home ranges do not appear to overlap, little intergroup aggression has been reported. The Ugandan population occupied relatively large home ranges, and failed to cover them daily. Rudran estimated their size as 72.5 hectares. It appears that this taxon is territorial (Table II).

Blue monkeys avoid chimpanzees, which are known to prey upon them. Their only interspecific interactions during the course of Rudran's study were aggressive in character. They fought with mangabey, redtail, red colobus, and black and white colobus monkeys. Subadults and juveniles aggressed more frequently against other species than expected by chance.

Blue monkeys feed on a wide variety of plants and invertebrate life. Rudran recorded at least 59 plant taxa and 101 specific food items for one group. Invertebrate consumption constituted nearly 20% of the total feeding observations. They were observed to consume butterflies, grasshoppers, ants, and cocoons with pupae. One subadult male was observed in the act of eating an African wood owl, which it presumably captured alive. Rudran speculated that the consumption of invertebrate food items by females may add needed protein during the gestation of young. The most important component of their diet is fruit, although foliar foods are also important seasonally.

Cords and Rowell [1986] studied a group of blue monkeys in western

TABLE II. Distribution of Various Forms of *C. mitis* in American Zoos[a]

C. mitis (n = 5)	
1.1	Toledo Zoological Gardens, Toledo, OH
0.2	San Francisco Zoological Gardens, CA
C. m. albogularis (n = 21)	
2.4	Brookfield Zoo, Chicago, IL
0.1	Central Florida Zoo, Sanford, FL
0.3.2	Cheyenne Mountain Zoological Park, CO
1.0	El Paso Zoological Gardens, TX
1.1.1	Greater Baton Rouge Zoo, LA
1.1	Monkey Jungle, Goulds, FL
1.2	The Zoo, Gulf Breeze, FL
C. m. kibonotensis (n = 1)	
1.0	Central Texas Zoo, Waco, TX
C. m. moloneyi (n = 4)	
2.2	Louisiana Purchase Zoo, Monroe, LA
C. m. stuhlmanni (n = 18)	
1.0	Ellen Trout Zoo, Lufkin, TX
2.2.1	Greater Baton Rouge Zoo, LA
1.2.1	Henry Doorly Zoological Gardens, Omaha, NB
1.1	Houston Zoological Gardens, TX
2.3.1	Louisiana Purchase Zoo, Monroe, LA

[a]According to the records of ISIS (as of June 30, 1986).

Kenya that was comprised of 42–48 members, excluding adult males. The females were typically permanent members, while males were transient. These investigators documented the fission of this large group into two ''daughter'' groups.

Cercopithecus mitis albogularis and Related Taxa

The closely related Sykes' monkeys are regarded by some taxonomists as differing from the blue monkey at the species level. They reside in lowland to montane forests with primary growth, and rarely come to the ground [Aldrich-Blake, 1970; Rowell, 1971]. They move largely in the forest canopy [Gartlan and Brain, 1968]. The home range of this species has been estimated to be 0.05 km^2. They have been observed to eat a greater portion of leaves than do vervet monkeys. According to Kingdon [1971], there are nine subspecies of Sykes' monkeys, all of which are identified by a white ruff.

Cercopithecus neglectus

This monkey is discontinuously distributed throughout equatorial Africa, more commonly in the western regions of the continent. Known as

TABLE III. Distribution of *C. neglectus* in American Zoos[a]

C. neglectus (n = 88)	
1.1	Zoo Atlanta, Atlanta, GA
1.2	Baltimore Zoo, MD
0.2	Jimmy Morgan Zoo, Birmingham, AL
2.1	Buffalo Zoological Gardens, NY
1.1.4	Busch Gardens, Tampa, FL
1.0.1	Central Texas Zoo, Waco, TX
2.3	Denver Zoological Gardens, CO
1.0	Duluth Zoo, MN
0.1	El Paso Zoological Gardens, TX
1.2	Glen Oak Zoo, Peoria, IL
1.1.1	Greater Baton Rouge Zoo, LA
1.1.1	Henry Doorly Zoo, Omaha, NB
1.1	Honolulu Zoo, HI
1.1	Houston Zoological Gardens, TX
3.2	Lincoln Park Zoo, Chicago, IL
1.1.3	Los Angeles Zoo, CA
0.1	Louisiana Purchase Zoo, Monroe, LA
1.1	Memphis Zoological Garden, TN
1.3	Monkey Jungle, Goulds, FL
1.2	New York Zoological Park, Bronx, NY
0.1	Philadelphia Zoological Garden, PA
1.1.2	Potawatomi Park Zoo, South Bend, IL
1.2	Potter Park Zoo, Lansing, MI
1.0	Racine Zoological Park, WI
1.1	Riverbanks Zoological Park, SC
0.2	Sacramento Zoo, CA
0.1	San Antonio Zoological Park, TX
2.4	St. Louis Zoological Park, MO
1.2	St. Paul Como Zoo, MN
2.1	Terry Lou Zoo, Scotch Plains, NJ
1.1	Utica Zoo, NY
1.2	Woodland Park Zoo, Seattle, WA

[a]According to the records of ISIS (as of June 30, 1986).

Debrazza's monkey, the animals live in forests along stream/river banks, preferring the mid-to-low canopy. They move quietly through their relatively small home ranges and show a marked sexual dimorphism with males averaging 7 kg and females 4 kg [Brennan and Else, 1984] (Table III).

Erythrocebus patas

Patas monkeys are primarily ground dwelling primates that inhabit northern equatorial woodland and grass savannahs, and wooded and grass

TABLE IV. Distribution of *E. patas* in American Zoos[a]

E. patas (n = 80)	
2.2	Abilene Zoological Gardens, TX
3.0	Birmingham's Jimmy Morgan Zoo, AL
2.3.1	Cheyenne Mountain Zoological Park, CO
1.0	Cohanzick Zoo, Bridgeton, NJ
1.2	Denver Zoological Gardens, CO
0.1	Greaer Baton Rouge Zoo, LA
1.0	Kansas City Zoological Gardens, MO
0.1	Louisville Zoological Garden, KY
0.0.1	Milwaukee Co. Zoological Park, WI
1.8	North Carolina Zoological Park
2.4.1	Racine Zoological Park, WI
1.1	Rock Island Co. Zoological Preserve, Moline, IL
0.1	San Diego Zoo, CA
1.2	San Diego Wild Animal Park, CA
1.7.4	San Francisco Zoological Gardens, CA
2.3	Staten Island Zoological Society, NY
1.5.5	Toledo Zoological Garden, OH
6.4	University of California, Berkeley, CA
1.2	Wildlife World Zoo, Glendale, AZ
2.5	Woodland Park Zoo, Seattle, WA

[a]According to the records of ISIS (as of June 30, 1986).

steppes that contain thickets [Hall, Boelkins, and Goswell, 1965]. Males weigh an average of 10 kilograms while females average 5.6 kilograms [Fedigan, 1982]. They are not reported to be territorial, as is typical of many open country cercopithecoids [Nagel and Kummer, 1974]. However, groups of patas seem to avoid each other and may be antagonistic when they meet [Fedigan, 1982]. Home ranges have been estimated at 51.8 km. [Hall, et al., 1965] and they typically move between 500–1200 meters per day. Group members disperse widely in the evening and sleep in scattered trees and bushes [Fedigan, 1982]. Precipitation in the areas that Patas inhabit is generally low and seasonal, but Gartlan [1974] reported that they drink 2 or 3 times per day if possible (Table IV).

Miopithecus talapoin

Talapoins are forest-dwelling primates that are annual breeders [Rowell and Richards, 1979]. The adult male averages 1.4 kg and the female 0.9 kg [Scruton and Herbert, 1970] (Table V).

TABLE V. Distribution of *M. talapoin* in American Zoos[a]

M. talapoin (n = 20)	
1.3	Henry Doorly Zoological Gardens, Omaha, NB
4.5	San Diego Zoo, CA
4.3	University of California, Berkeley, CA

[a]According to the records of ISIS (as of June 30, 1986).

SOCIAL ORGANIZATIONAL FACTORS

Cercopithecus aethiops

Vervets generally live in multimale, multifemale troops with reported sizes ranging from 7 to 76 animals [Fedigan, 1982; Kavanaugh, 1980] and an average of 12 [Hall and Gartlan, 1965; Dunbar, 1974] to 24 animals [Struhsaker, 1967]. At least one study has reported unimale bands ranging from two to 19 members [Struhsaker, 1967]. Poirier [1972] describes a typical group as containing two adult males, four adult females, four subadults, six juveniles, and four infants. Matrilineal groupings are important influences on social interactions [Fedigan, 1982]. More dominant lineages are reported to be more cohesive [Ehardt-Seward and Bramblett, 1980].

Group movements are initiated in approximately equal frequency by adult males and adult females [Fedigan, 1982]. The adult males have a strong vigilance role and often sit in exposed trees or rocks, apparently acting as "sentries." Rocks are also used as resting areas at times [Poirier, 1972]. Objects such as these ought to be available in captive settings. Group dispersal is often significant during the day and at night [Hall and Gartlan, 1965], although Dunbar [1974] reported that green monkey groups usually remain spatially compact.

The pattern of male migration seems to vary among different areas. Solitary or peripheral males are reported in some cases, especially where troop density is low, and a small number of all male groups have been noted [Fedigan, 1982]. Male migrations coincide with the mating season and many males migrate more than once [Henzi and Lucas, 1980]. No alpha male was reported to move until after he had lost his status. Orphans transferring with their brothers or peers are common migrants [Cheney et al, 1981]. Fairbanks and McGuire [1984] used this information in determining their management practices in an outdoor colony of vervets. Natal males were removed from their groups between 4 and 6 years of age, and alpha males were removed after tenures of 3 to 5 years.

Cercopithecus mitis stuhlmanni

Rudran [1978] reported group sizes of 13 to 27 animals. Four of the five groups he observed contained one adult male and the fifth group contained two adult males. The groups ranged in size from four to 12 adult females and seven to 16 immature members. Some solitary animals were also observed. Young males were reported to emigrate from their troops.

Cercopithecus mitis albogularis

There have been no field studies of the Sykes' monkey in which natural social organization has been reported. Rowell [1971] discovered that in one captive group there was a dominance structure, and that the male was *not* the highest ranking animal. Most social interactions involved sitting together. Tail-twining occurred during some of this time and also when slight disturbances arose.

Cercopithecus neglectus

The Debrazza's monkey is apparently a monogamous primate over some of its range, a rare phenomenon among the cercopithecines. They associate in small social groups [Brennan and Else, 1984].

In captivity, Brennan (in press) found that 43.7% of the world's breeding units were monogamous housing situations.

Erythrocebus patas

Patas are reported to live in two types of social groupings—unimale harem groups and nonreproductive all-male groups. Such all-male groups are typical of many open-country primate species [Gartlan, 1974]. Group sizes of four to 31 have been reported in the field [Hall et al, 1965], with an average of about 15 [Hall, 1966, cited in Preston et al, 1970], containing between four to 12 adult females [Fedigan, 1982]. The heterosexual groups appear to be dominant to the all-male groups, as they have priority at water holes [Gartlan, 1974] and extra-group males are often threatened and chased by members of the heterosexual group [Kaplan and Zucker, 1980].

The role of the adult male in reproductive groups appears to be largely one of vigilance, watching for both predators and other male patas [Fedigan, 1982]. The harem male is spatially and socially peripheral to the remainder of the group in both captive and wild settings [Gartlan, 1974; Nagel and Kummer, 1974]. The adult females are the more cohesive force in the groups, displaying the majority of grooming, interfering in fights, and other affiliative physical contacts [Kaplan and Zucker, 1980]. Females also seem to initiate and direct most group movements [Gartlan, 1974]. The members of an all-male group are not necessarily excluded from heterosexual life indefinitely. Many of them are young or wounded and it is expected that at some point they may become integrated into reproductive groups [Gartlan, 1974]. However, this process has not been documented in field studies [Fedigan, 1982]. Recently, Harding and Olson [1986] described temporary integration during the breeding season. Males typically leave their natal troop when they are between 1.5 and 4.5 years old. The young males often leave voluntarily, although at times they may be driven out by other group members [Gartlan, 1974]. Typically the resident harem male is the only one in the group with the bright blue scrotum and red perianal skin indicative of sexual maturation [Gartlan, 1974]. An increase in aggression directed by an adult male toward a maturing male was reported by Hall and Mayer [1967]. This behavior can be expected in captivity if the maturing male is not removed at an appropriate time.

Patas monkeys seem to require a large intergroup distance. In fact, Kaplan and Zucker [1980] suggested that a 94-acre island may not be large enough to support two heterosexual patas groups. A high degree of dispersal appears to be the most important means of social buffering within groups as well. Patas display a relatively linear dominance hierarchy within heterosexual groups with few reversals [Kaplan and Zucker, 1980]. This hierarchy does *not* seem to be an important influence on establishing affiliative relations through grooming, or in controlling scarce resources such as food. Affiliative behaviors such as grooming, interfering in fights, nose-to-nose greetings, and other physical body contact are relatively infrequent in patas monkeys [Kaplan and Zucker, 1980].

Miopithecus talapoin

In Cameroon, Rowell and Dixson [1975] observed troops composed of 80–92 talapoins. Gautier-Hion [1966, as cited in Aldrich-Blake, 1970] reported groups of 60 to 158 animals. Male group transfer may be rare because of the large group size [Scruton and Herbert, 1972]. Before the breeding season, adults resided in separate sex groupings. However, at the peak of the mating season many females joined the predominantly male subgroups and left their infants with the predominantly female subgroup [Rowell and Dixson, 1975]. Dominance hierarchies exist in which females often rank above the males [Rowell and Dixson, 1975; Wolfheim, 1977a]. Adult males tend to avoid adult females and associate mainly with each other. They are less active socially and less active overall than are the females [Wolfheim, 1977b]. At night the troop may divide such that individual females sleep with their young infants and infants from the previous year, while the males sleep separated from the females and young [Aldrich-Blake, 1970].

REPRODUCTION

Cercopithecus aethiops

Sexual behavior. Copulation seems to be relatively infrequent in vervet monkeys. In one captive study, only five heterosexual mounts were observed in 31 months; apparently four of these led to births [Bramblett et al, 1975]. The authors observed that the typical sexual interaction began with the male inspecting the anogenital region of the female. If she was in an estrous condition, copulation occurred. If the female was not receptive she either ignored or threatened the male. They also noted that the frequency of brief male mounts increased after caretakers intervened to swab females. A field study reported that males and females initiated copulations in approximately equal frequencies [Gartlan, 1969], and some females mounted the male if he rejected their approach. Gartlan [1969] also described a crouch by the female indicative of her receptivity. These females also sexually presented to males. Keddy [1986] found that female vervets preferred to mate with the alpha male under both dyadic and group living conditions. More dominant females were more successful at rejecting the sexual solicitations of more subordinate males.

Vervets display a single mount to ejaculation pattern with a mean of 9.5 thrusts per mount and a range of 3 to 15 [Gartlan, 1969]. The male clasps the female's legs with his feet and holds onto her sides during copulation. Each copulation is followed by a fairly long refractory period and only one female was seen to copulate more than one time per day. Copulations often occurred within view of other males, but they did not react to it, although juveniles interfered with 9.5% of the mounts [Gartlan, 1969; Schapiro, 1984].

Seasonality. Field research has found vervets to be reproductively seasonal [Bramblett, et al, 1975]. Every month of the year is a part of the birth season of at least one group of vervets. This has led some researchers to speculate that their synchrony is troop specific and may not be strongly determined by climate [Bramblett, et al, 1975]. In captivity, some seasonality is present, but at the same time some conceptions occur every month [Hess et al, 1979; Valerio et al, 1972]. Some investigators have found no evidence of a breeding season in captivity [Mallinson, 1971]. It seems that under some conditions, vervets are able to "break out" of seasonal breeding and/or birth patterns [Fairbanks and McGuire, 1984; Schapiro, 1984].

Else et al [1986] after 5 years of monitoring hormonal levels in 28 females, concluded that vervets are "marginally" seasonal breeders. The incidence of prolonged menstrual cycles (over 50 days in length) was reduced during the natural breeding period. Twenty percent of the recorded cycles were classified as prolonged.

Cycling. Female vervets reach menarche at approximately 30 months of age [Bramblett et al, 1975]. Females' cycles are 30 to 31 days in length [Hess et al, 1979; Johnson et al, 1973], with menses ranging from 1 to 12 days with an average of 3 to 4 [Hess et al, 1979; Valerio et al, 1972]. Most researchers classify cycles of longer than 50 days as amenorrheic [Hess et al, 1979; Valerio et al, 1972]. Female vervets show little visible vaginal discharge [Bramblett et al, 1979], and there is no sexual swelling [Gartlan, 1969]. It generally takes a period of 6 to 12 months of adaptation to captive life before cycles occur regularly [Valerio et al, 1972]. There is no pattern of male behavior that is consistently related to the estrous state of the female, and vervet females have been said to be in a state of "permanent estrus" [Rowell, 1970, 1971].

Breeding success. Vervets breed well in captivity. Their fecundity is only surpassed by smaller primate taxa that have multiple births [Else, 1981]. Female vervets can give birth to their first infant as early as 3 years of age, which is about 1 year earlier than the rhesus monkey [Rowell and Richards, 1979]. Gestation lengths have averaged between 157 and 185 days, with a modal report of about 165 days [Bramblett et al, 1975; Hess et al, 1970; Mallinson, 1971; Valerio et al, 1972]. Pregnancy can be determined by palpation at 25 to 30 days after breeding [Hess et al, 1979; Johnson et al, 1973]. Implantation bleeding is seen in approximatly one third of the pregnancies for 1 to 5 days in the second and third postbreeding weeks [Hess et al, 1970; Valerio et al, 1972]. The interbirth intervals range from about 320 to 344 days [Bramblett et al, 1975; Bramblett and Bramblett, 1982; Fairbanks and McGuire, 1984].

The conception rates of vervets are very high. In a controlled laboratory setting, Hess et al [1979] reported a conception rate of 21.8% for a single mating 2 days before midcycle, and 45.8% for multiple matings over a period of 2 to 3 days at midcycle. Valerio et al [1972] determined that 75% of their females conceived in a 1-year period, requiring an average of five matings. These investigators placed the female into the male's cage for 6 days on the tenth day following the onset of menses. Gartlan [1969] reported a reproductive success of more than 80% in the field, but even higher rates have been achieved in captivity. Of 109 females, 92% became pregnant and 89% gave birth within 1 colony year in Philadelphia [Kusher et al, 1982].

Fairbanks and McGuire [1984] reported that 22% of the recorded pregnancies in their colony ended in miscarriages or stillbirths, and 13% of the infants died before 1 month of age. Another 6% died between 1 and 13 months of age in one colony. Mortality rates are higher for infants with mothers younger than 6. There is no evidence that shortening the interbirth interval reduced the probability of survival of the nursing infant. Else [1981] obtained 90% live births annually. Twenty-five of their 26 breeding groups in Nairobi produced a live born infant within 1 year of the initiation of the colony. At their large colony in Sepulveda, California, Fairbanks and McGuire [1984] obtained a mean interbirth interval of 10.7 months with a median of 10.2 months. Thus it seems that vervets consistently produce infants at a faster rate than macaques. It is important to note that this increase in fecundity did not lead to a decrease in infant survival. Kushner et al [1982] estimated that 180 females could produce 100 offspring that survive to maturity annually.

Similarly, Valerio et al [1972] reported a 28% fetal wastage rate. Pregnancy can be a serious health problem and five females in this colony died due to organ failure, hemorrhage, or unsuccessful delivery attempts. The miscarriage and stillbirth rates are high when compared to field data and the authors speculated that this may be due to stress, diet, and other factors associated with confinement. Johnson et al [1973] suggested that the excitability of vervets may influence the high abortion rates. Postpartum menses occurs an average of 60 to 84 days following parturition [Hess et al, 1979], and copulation often occurs within that first month [Mallinson, 1971; Fairbanks and McGuire, 1984].

Social organization to facilitate breeding. From the previous section it is apparent that the highest rates of productivity are reached when males and females are housed together continuously. The Fairbanks and McGuire group live in a large (350 m^2) outdoor enclosure housing 21 adult females and five adult males; they reported very impressive rates of production. However, the Kushner et al [1982] group also had very high production rates; these breeding groups contained one male and two to six females that lived in outdoor housing units measuring 4 × 5 × 7 m with exercise areas of 5 × 18 × 7 m. In this case, pregnant females were removed before the time of their delivery and were returned with their infant to the group 1 to 2 weeks postpartum. Apparently, vervets can reproduce well under extremely diverse conditions.

Cercopithecus mitis stuhlmanni

Very little information concerning *C. mitis* reproduction is available in the literature. Rudran [1978] observed that their births are seasonal in the Kibale Forest of Uganda, with December through February being the birth months.

During mating periods, extra-group males may join the longer term resident male. Tsingalia and Rowell [1984] recorded 18 such males joining an established group during the breeding season. There was a strong suggestion that copulations were more frequent when more than one male was present. Captive breeding might be improved by exposure to additional males.

Cercopithecus mitis albogularis

Reproduction of the Sykes' monkey has been more fully documented than that of the blue monkey and may serve as a useful model for blue monkey reproduction.

Sexual behavior. Sykes' monkeys show relatively low levels of sexual behavior [Schapiro, 1984]; it is unusual to observe more than one copulation per hr in a captive setting [Rowell, 1971]. The sexual interactions are brief and are initiated by both males and females [Rowell, 1971]. Estrous females show no sexual swelling, but they may inspect and nuzzle the genitals of the male [Bramblett et al, 1975] and occasionally may present their hindquarters to the male [Rowell, 1971]. The male may also exhibit similar nuzzling and inspecting behaviors. The male typically mounts three or four times before intromission is achieved, and he may chase and threaten juvenile animals between these mounts. If there is no harassment of the mating pair, ejaculation is usually achieved on the first mount [Rowell, 1971].

Seasonality. Some lab researchers claim that Sykes' monkeys do not reproduce seasonally or display synchronized estrous cycling [Else et al, 1985; Schapiro, 1984]. However, Rowell [1970] reported that wild Sykes' have one breeding season and that captive Sykes in Uganda had two breeding seasons annually.

Cycling. Female Sykes' mature later than do vervet monkeys [Rowell and Richards, 1979]. Their average cycle length is 30 days [Rowell, 1970] and menses is generally not visible. Sykes' apparently copulate within an estrous period but not at a consistent stage of the cycle, so it is not clearly defined [Rowell, 1970]. A mean cycle length of 31.9 days was derived from the examination of 59 cycles for five females of the *Kolbi* subspecies. The mean menses length during these cycles was 5.3 days.

Breeding success. Rowell and Richards [1979] conducted the most thorough analysis of Sykes' reproduction and concluded that Sykes' are later maturing and slower breeding than many monkeys; for those reasons they would not make good candidates for general purpose laboratory animals. The authors found that the median age at first parturition was approximately 5½ years. This is older than the corresponding age for vervets, patas, and DeBrazza's monkeys all housed in the same place. Rowell [1970] recorded an average gestation of 140 days. Interbirth intervals averaged 413 days [Bramblett et al, 1975], but several abortions affected that measure. Rowell and Richards [1979] reported that 13 Sykes' mothers gave birth to 46 infants over a 14-year period. Some fetal loss may be undetected since abortions are consumed [Bramblett et al, 1975], and relatively high abortion rates may be expected as a new colony of animals are adjusting to caging [Bramblett et al, 1975]. Female Sykes' may remain active breeders for longer than other related monkey species. Two females were still breeding when they were over 14 years of age, and one was over 20 years old [Rowell and Richards, 1979]. In comparison, a 12-year-old female vervet is generally beyond the age of reproduction.

Else et al [1985] recorded 35 live births and 11 stillbirths from 21 of 25 female blue and Sykes' monkeys over a period of approximately 5 years. The females gave birth at least twice with a mean interbirth interval of 19.2 months. This is considerably longer than the 15-month interval reported for wild populations [Butynski, 1982]. The stillbirths constituted 24% of the total births. Abortions could not be detected.

Social organization to facilitate breeding. It may be important to allow infant Sykes' monkeys to leave the enclosure that houses adult animals to maximize production. Rowell [1970] reported that conception may be delayed by housing infants with their mothers beyond 5 months of age. When infants were restrained in the adult cage, one adult female did not conceive, although she did copulate. Another female conceived but rejected her newborn offspring and nursed her older infant instead. When infants were allowed to leave the adult cage, one female gave birth only 8 months after her first infant was born.

Blue monkeys are apparently not amenable to timed matings, since 45 attempted pairings involving six females resulted in no conceptions in one laboratory [Else et al, 1985]. This was true even though sperm were present after 35 of those 45 matings, and the mating period spanned what is believed to be the optimal period for conception.

Else et al, [1985] concluded that while both blue and Sykes' monkeys do not exhibit overt stress reactions to captivity, their menstrual and hormonal patterns indicate that several years are required for reproductive adjustment to a captive life.

Erythrocebus patas

Sexual behavior. Female patas monkeys do not show any external signs of sexual receptivity such as sexual swelling but they do engage in an unusual form of "cringing" locomotion that is apparently a reliable indicator of receptivity [Hall et al, 1965; Hall and Mayer, 1967]. This cringing movement is exhibited along with a toneless sound made from forcing air into the cheek pouches [see Hall et al, 1965, for a complete description] and idiosyncratic forms of masturbation [Rowell and Hartwell, 1978]. The male often approaches a female after she displays the sexual solicitation described above. Males and females solicit approximately an equal number of mounts, but males reject more female initiations than the reverse [Hall and Mayer, 1967; Rowell and Hartwell, 1978].

The male usually approaches the female from the rear, inspects her anogenital area, and may lift her to a quadrupedal position. The female may run away at this point, terminating the interaction. Sometimes, however, the female may cringe away and the male may follow her. The male does not use a foot clasp while mounting, and repeated mounts are usually performed before ejaculation occurs. Following a mount the male often sits down and grooms his penis [Hall et al, 1965]. There is some indication that proper sexual behavior requires experience, as a young male has been observed to mount unreceptive females and then to ignore a soliciting female [Hall and Mayer, 1967].

Seasonality. There is strong evidence for seasonality in breeding and births among wild patas monkeys [Hall et al, 1965; Rowell and Hartwell, 1978]. The situation is not quite so clear in captive settings. Some researchers have obtained evidence for seasonality [Kaplan et al, 1981], but others have housed patas without finding any evidence for seasonality [Rowell and Richards, 1979]. One researcher reported a tendency for fall mating, but found sperm in vaginal samples during all months of the year as well as during pregnancy [Hartwell, 1978]. Rowell and Hartwell [1978] found that several females were usually receptive simultaneously.

Breeding success. Patas reproduce well in captivity [Chism et al, 1978; Rowell and Richards, 1979]. Since female patas can produce their first infants approximately 1 year earlier than rhesus, and since they have a median birth interval that is about 2 months shorter, they are considered to be a good laboratory primate. Conception rates are quite high. In one facility, none of the females had to be mated more than twice to conceive [Loy et al, 1978]. Rowell and Hartwell [1978] reported 16 livebirths out of 26 conceptions, and in another colony six pregnancies led to five healthy infants [Loy et al, 1978].

Gestation lengths of 164 [Hartwell, 1978] to 168 days [Loy et al, 1978] have been calculated in captivity, and these generally correspond to those estimated in the field [Loy et al, 1978]. Patas have an interbirth interval of about 60 weeks with about 37 weeks between a birth and the next conception [Rowell and Hartwell, 1978] and 22 weeks from a birth to the next menstruation [Loy et al, 1978]. Longer intervals may be expected from lower ranking females having their first or second infant [Loy et al, 1978], and more variable intervals are recorded following stillbirths [Rowell and Hartwell, 1978]. The female's postpartum return to sexual receptivity seems to be influenced by her social environment. Whether the postpartum females were lactating or not, they became receptive sooner when the rest of their group was reproductively active [Rowell and Hartwell, 1978]. The presence of infants, however, may reduce the group's sexual behavior.

Social organization to facilitate breeding. Kaplan et al [1981] compared the effectiveness of two different strategies for housing and breeding patas monkeys. The more traditional method of housing one male continually with four or five females and their offspring was shown to be inferior to the daily alternation of one adult male between two groups of females and the removal of pregnant females. With the new information available from at least one field site, it may be advantageous to simulate extra-group males joining the group to participate in mating during the breeding season [Harding and Olson, 1985]. This may increase aggression and wounding but contribute to genetic diversity. Juvenile animals were placed in peer groups or removed for experimental purposes. Groups of patas housed adjacently were rearranged so that groups of vervet monkeys separated them. The authors concluded that this buffer reduced the number of male aggressive displays. This method also led to an increased percentage of females having viable offspring from 35–43% up to 80%. The females were placed in individual cages when they become pregnant and were integrated into female-offspring groups approximately 4 to 6 weeks after parturition. The authors claimed that a reduction in the kidnapping of infants was an important factor in decreasing the number of neonatal deaths.

While earlier field studies reported a single resident adult male and a resultant harem mating system, Harding and Olson [1986] observed another system at a Kenya field site. An influx of extra-group males mated within the group during the breeding season. There were as many as six males in the group simultaneously, with a total of ten extra-group males mating with the resident females during the 10 weeks of conception. The males displayed a combination of consort behaviors and opportunistic mating strategies. Harding and Olson reported that few conceptions resulted from matings when the single resident male was present in the group, as compared to the greater number when many males were present.

Miopithecus talapoin

Sexual behavior. According to Rowell and Dixon [1975], researchers describe the courtship behavior of talapoins as particularly elaborate. This has been attributed to the males' fear of possible aggression from the females. Completed copulations seem to be preceded by less courtship in wild conditions than in captive conditions, as free-ranging males were observed to mount a female after only one or two visual inspections of her perineum. The female may initiate courtship with sexual presentation, or the male may initiate it with genital inspections and following the female. Single mounts may end in ejaculation, though two or three mounts are sometimes observed before ejaculation. Both sexes may emit high pitched screeching during mating [Scruton and Herbert, 1970]. The interejaculatory interval for males is 45–60 min, and receptive females return to courtship 5–10 min after copulating [Rowell and Dixson, 1975]. Scruton and Herbert [1970] reported that the lower ranking of the two males in the talapoin group they studied displayed little or no sexual contact with any of the females.

Seasonality. In field studies, the talapoin breeding season lasts just over 2 months [Rowell and Dixson, 1975]. This is shorter in duration than the breeding season in captive groups [Rowell and Dixson, 1975], and may not begin as abruptly as the breeding season does under more natural conditions [Rowell, 1977]. Seventy percent of the conceptions in one captive talapoin population occurred between December and March with no conceptions between June and September [Rowell, 1977].

Cycling. Talapoins have a relatively long adolescence, with swellings of their sexual skin not beginning until the fourth breeding season [Rowell, 1977]. The mean cycle length in talapoins is reported as 32.9 days in one laboratory study [Scruton and Herbert, 1970] and 35 days in another [Rowell, 1977]. Menstruation usually occurs when the sexual skin is deturgescent. The stage of maximum swelling is 2–6 days in duration and deflation then occurs until a few days before menstruation. Scruton and Herbert [1970] state that a few females had no swelling of the sexual skin at any stage of their cycle. Rowell [1977] reported that the female skin color grows darker and shinier as sexual activity increases, at times within a period of only 15 min.

Breeding success. Rowell and Dixson [1975] discovered that the majority of the 40 females in the two wild troops they studied were probably monoestrous. Talapoins are apparently so fertile that completed menstrual cycles are rare [Rowell, 1977].

Scruton and Herbert [1970] recorded a gestation of approximately 158 days for a live birth and Rowell [1977] reports gestations of 158 to 166 days. Of 27 pregnancies, 11 led to livebirths, seven were full term stillbirths, and nine ended in abortions [Rowell, 1977]. Six of these nine abortions occurred during the second one third of the pregnancy. Aborted fetuses were generally eaten by the mother if the fetus did not have hair. Following a stillbirth, intervals of 29–75 days passed before the female returned to cycling [Rowell, 1977]. Following the death of infants that were 25 and 44 days of age, the mothers began cycling between 14 and 17 days later.

Social organization to facilitate breeding. In the wild, adult talapoins live in single-sex subgroups during nonbreeding seasons. The males begin to move into the female subgroups before visual signs of perineal swellings are obvious [Rowell and Dixson, 1975]. Males frequently returned to the all-male groups directly after engaging in sexual behavior. The single-sex subgroups reformed after the breeding season. Rowell and Dixson [1975] believe that this regrouping may be at least partially stimulated by the increased aggressiveness of the newly pregnant females. It may be important to allow these subgroup changes in captivity to most accurately simulate wild conditions. Rowell and Dixson [1975] believe that the breeding season in the wild is shorter in duration and involves a greater degree of social disruption than in captive groups. This disruption may include chasing, threatening, and mobbing [Rowell, 1977; Rowell and Dixson, 1975].

Wolfheim [1977a] reported that young male talapoins were killed by females during the breeding season. Wolfheim believes that the lack of opportunity to distance males from females after the breeding season may have led to the deaths of these male talapoins.

BIRTH, PARENTING, AND INFANT DEVELOPMENT

Cercopithecus aethiops

Vervets normally give birth at night [Mallingson, 1971]. Fairbanks and McGuire [1984] note that in each of the two cases where daytime parturitions were observed, the mothers died. One normal labor that was observed lasted approximately 1 hour [Brandt and Mitchell, 1971]. Following birth the mother generally eats the placenta [Gartlan, 1969; Johnson et al, 1973], cleans the infant, and may crouch over it [Gartlan, 1969]. Females with infants remain in particularly close association with one another.

Adult male vervets seem to show no interest in vervet infants [Gartlan, 1969], but simply tolerate their contact. Adult males interact more with juvenile aged animals [Fedigan, 1982].

Intense allomothering is reported with vervet subjects. Infants may be passed among females beginning at 2 to 14 days of age [Gartlan, 1969] but Lancaster [1971] reports no passing of infants. Subadult females commonly serve as "aunts" to infants [Gartlan, 1969; Johnson et al, 1980; Lancaster, 1971; Mallinson, 1971]. These young females are sometimes clumsy with the infants, and may retain the infants tenaciously [Gartlan, 1969]. Allomothering may be dangerous to infants, as two of four captive infants that died during a 1-year period in one lab had empty digestive tracts. Both of those infants were extensively allomothered [Schapiro, 1984].

While allomothering may have negative consequences, it may also be beneficial to the infant and allomother. Mallinson [1971] discovered that a lack of aunting behavior led to severe "fits" by unattended captive vervet infants. Of six infants observed, two that had animals other than their mother with which to huddle did not display these fits. Four of the infants did throw fits when their mother left them alone, often while she groomed her adult male partner. These infants would collapse on the cage floor with limbs jerking, eyes static, and bodies eventually becoming motionless until the mother was "sufficiently stimulated by outside human aggression to gather the infant safely up" [Mallinson, 1971, p. 602]. To overcome this problem, the mother-infant pairs were placed alone in a strange environment for 1 month. The mother became more possessive during this time, and the pairs were subsequently reintroduced into the cage with the male or males. The authors suggested that it may be necessary to have a female available for aunting to achieve successful rearing without human intervention.

Vervet infants develop faster than other *Cercopithecus* monkeys [Fedigan, 1982]. The newborn can cling to its mother soon after birth and can ride ventrally unsupported by the time it is a few weeks old [Gartlan, 1969]. The infant may nibble at food for the first time between 3 and 12 days of age [Mallinson, 1971]. Mallinson [1971] suggested that the captive group be fed to saturation to ensure leftover food for the infant (as would be available in the wild). At about 3 months of age the infant's coat changes from black to light gray in color. At about this time the mother begins to refuse to carry the infant, unless some danger is present [Lancaster, 1971]. The infant may begin to enter a juvenile play group and form social bonds with other group members. Captive infants may be weaned at 6 months of age [Kushner et al, 1982]. In one vervet colony [Kushner et al, 1982] the weanlings are segregated into peer groups by age and sex. Group members of other ages may, however, be important for social development. Fedigan [1972] found that captive adult female vervets reinforced the play of immature monkeys and adult males played roughly with the older juvenile males. Juvenile animals stimulated play in young infants. Perhaps mixed-age groupings would encourage adequate social development.

Few vervet females in Kenya's Amboseli National Park gave birth in consecutive years, and many suckled their infants into the second year of life. There is evidence that stressed females weaned their infants relatively early [Lee, 1984].

Cercopithecus mitis

Rudran [1978] reported that adult females with infants in the wild form subgroups that remain more distant from other animals.

Cercopithecus mitis albogularis

Sykes' infants develop more slowly than vervet infants. Sykes' infants do not break contact with their mothers within their first 4 weeks of life and they have a relatively long nursing period [Rowell, 1970]. Rowell [1970] discovered a reversion to higher levels of clinging and suckling when infants were kept with their mothers beyond the age of 5 months compared to a housing situation in which infants were able to leave the adult's cage.

Cercopithecus neglectus

Stevenson [1973] described births in two DeBrazza's females. Females tightly cradled the newborn infants and each infant suckled soon after birth. The infants first moved away from their mothers at between 5 and 7 days of age. At approximately 50 days of age the infants began to show interest in solid food. When the infants were about 2 months old they began engaging in social play.

The adult male was passive toward infants in general, but occasionally would play bite the infants. The male first threatened an infant when it was over 4 months old.

Brennan (in press) surveyed a population of 575 captive DeBrazza's monkeys in 81 institutions worldwide. She concluded that a high level of infant mortality and a low level of fertility is a clear threat to the captive population.

Erythrocebus patas

There is little abdominal swelling associated with pregnancy until only a few weeks before parturition [Goswell and Gartlan, 1965]. Although many primates typically give birth at night, patas seem to be more likely to give birth during daylight hours [Chism et al, 1978]. Early signs of an ensuing birth include mucous and blood discharge from the vagina, lying down, feeling the vulva, crouching, and lifting the tail as if to defecate [see Goswell and Gartlan, 1965, for a complete birth description]. These behaviors are seen 1 to 8 hours before the birth occurs, although a bloody discharge has been noted even 10 days before birth [Goswell and Gartlan, 1965]. The mother may or may not eat the placenta [Chism et al, 1978; Richman, 1963]. Some experienced mothers have been known to put their newborn infants down while they eat the placenta [Chism et al, 1978]. The mother often cleans her own hands and birth area before she cleans the infant.

Some reports indicate that patas mothers do *not* avoid other group members at the time of parturition, and they may even prefer to be near them [Chism et al, 1978]. Rowell and Hartwell [1978] found an increase in both "friendly" and avoiding behaviors following a birth. The new mother may be groomed more, and the birth may cause a decrease in female-female aggression [Hall and Mayer, 1967]. For these reasons it may be advisable to avoid isolating new patas mothers.

The adult male does not interact with infants, but the mother and infant form a relatively brief but intense bond [Kaplan and Zucker, 1980]. Following birth the mother lifts the infant to her teat to begin feeding and assists in maintaining ventro-ventral contact. Patas mother-infant pairs interact more during the first 6 months of the infant's life than do macaques [Preston et al, 1970]. Seven-month-old patas exhibit a typical though short-lived reaction to separation from their mother. Surprisingly, the infants showed an increase in *peer* interaction following reunion with their mothers, rather than a sustained increase in maternal interaction [Preston et al, 1970]. The authors suggested that patas mothers may inhibit this peer play by threatening other infants.

Patas are relatively fast maturing primates [Rowell and Richards, 1979], and the infants are precocious. Infants grasp their mothers' hair immediately following birth, cling unaided to their mother by the time they are 1 hour old [Chism et al, 1978], and they can locate the teat by this time as well [Hall et al, 1965; Goswell and Gartlan, 1965]. By 2 days an infant may try to leave its mother's grasp [Richman, 1963], and by a few days the infant rarely sits in her arms except when suckling [Hall et al, 1965]. By 1 week of age the infant creeps away from the mother and begins to grasp and bite at objects, although it quickly returns to its mother at any sign of danger. Play is common among young patas, usually consisting of chasing and bounding off bushes [Preston et al, 1970]. They do not engage in as much contact play with peers as do macaques, but they play more with their mothers than do young macaques. Object play is also commonly seen among young patas [Hall et al, 1965].

Although adult male patas do not interact with infants, the other females in the group are attracted to neonates [Hall and Mayer, 1967; Zucker and Kaplan, 1981]. Zucker and Kaplan [1981] found that adult females exhibited the vast majority of the allomaternal behavior, consisting largely of contacting and nuzzling the infant. They also observed eight instances of kidnapping, none of which led to any type of agonistic encounters. The authors suggested that allomaternal behavior can be considered an integral part of patas social life. However, in captivity, kidnapping may be more harmful. Kaplan et al, [1981] reported several infant deaths apparently caused by kidnapping with a failure to return the infant to its mother. Hall and Mayer [1967] found that a patas mother repelled attempts to contact her infant for about 2 weeks, but once the initial allomaternal contact occurred, the mother no longer tried to prevent it. It is important to note that this mother seemed able to retrieve her infant at any time. Kaplan et al [1981] suggested that in field conditions kidnappings are shorter in duration, and mothers can easily retrieve their infants. In the captive situation, mothers may not be able to avoid would-be kidnappers because of the lack of space, and this may cause the less favorable outcomes of some captive kidnappings. Opportunities to hide from other group members (e.g. "cover") may reduce this problem.

Patas exhibit the fastest rate of sexual maturation among the Old World monkeys [Chism, 1986]. By 5 weeks patas infants can walk steadily, and they begin to exhibit social play approximately 3 weeks later [Chism, 1986]. Chism also reported that at a median age of 9 weeks, male and female patas exhibited mounting in play and exploratory contexts. Mounting was only seen in sexual contexts among adult patas.

In Chism's study group weaning began by 12 weeks old, as patas mothers began rejecting nursing attempts of their infants. However, these rejections, and punishments in general, were quite infrequent and mild during the infants' first 3 months, (less than one nursing rejection per 5 hr of observation; one other rejection per 10 hr of observation). Beyond 5 months of age, patas infants spend less than 5% of their time nursing or resting in contact with their mothers. Nursing was continued generally until the mother gave birth again. This was usually at the end of 1 year, but nursing continued during the infant's second year if a subsequent birth did not take place until then.

After about 9 months of age, the young patas males began displaying harassment of adult males. This harassment consisted of approaching and following the adult male while squealing and hitting and grabbing the adult.

The harassment took place only when group females were sexually receptive. The infant and juvenile males combined to harass every observed copulation attempt in this study.

By 1 year of age, patas youngsters were independent members of their social groups and functionally exhibited at least rudimentary forms of all adult social behaviors except for adult female sexual behavior.

Chism [1986] observed some intense maternal resistance to allomothering transfer attempts and noted that maternal rank did not influence whether or not infants were transferred. However, high-ranking mothers may have been able to retrieve their infants with greater ease. At 2 weeks of age, infants spent 25% of their waking time being carried by an allomother. However, by the time infants were 3 months of age, this allomothering was no longer observed [Chism, 1986]. Some infants continued to have close associations with females that had frequently been allomothers to them. Chism [1986] noted that the lack of allomothers may lead to slower rates of social development and higher levels of maternal protectiveness.

There are no systematic accounts of hand-rearing techniques used for patas monkeys; however, some brief mention of success or failure is made in a few publications. Goswell and Gartlan [1965] reported that they successfully hand-reared a 6-week-old infant whose mother's milk had dried up. Loy et al [1978] attempted to rear an infant born about 2 weeks earlier than average, but they were not successful. Kaplan and associates [1981] reported that two of their four attempts at hand-rearing were successful.

Miopithecus talapoin

With pregnancy, the female talapoins' sexual skin deflates, although it may be somewhat reddened or slightly swollen [Scruton and Herbert, 1970; Rowell, 1977]. Infants weigh about 200 gm [Rowell, 1977] which is about one fifth of the mother's weight. The infants are relatively slow maturing, with lactation lasting between 49 and 210 days [Rowell, 1977]. Hill [1966] described a single infant's development. The mother first left the infant for 15 min when it was 7 days old. At 16 days old, the infant attempted to eat food, and was largely weaned by 50 days of age. However, the infant continued to attach to the nipple when it was asleep. No weaning tantrums were observed. Brandt and Mitchell [1971] described one talapoin birth in which the female ate the placenta and placed her infant on the floor on its ventrum when the infant was only 2 hr old. The mother hovered over the infant, tightly pressed it to the floor with her hands, and thrust against it 6 to 8 times. She then retrieved and groomed the infant.

AGGRESSION

Cercopithecus aethiops

Vervet dominance relations do not seem to be as clear-cut as they are in macaques or baboons. Male and female dominance hierarchies appear to operate independently, with some females dominant over some males and vice versa [Lancaster, 1971]. The adult male in one captive group was chased by all of the females and was not avoided by them [Rowell, 1971]. The tenure of three alpha males ranges from 8 to 30 months [Henzi and Lucas, 1980]. Social interactions are generally peaceful, with sitting together and grooming being the most frequent mode of interaction [Rowell, 1971]. Increased and more intense aggression has also been recorded during the breeding season [Henzi and Lucas, 1980]. Coalitions are important in aggressive encounters [Lancaster, 1971; Struhsaker, 1967]. Fedigan [1982] reported that maternal defense is a frequent basis for female coalitions. Rowell [1971] claimed that high-ranking animals intervened in agonistic encounters. Both male or female animals may interfere in this manner. Vervets may also defend group members against a human caretaker [Fedigan, 1982]. Amboseli vervets exhibited reduced levels of play and aggression when food quality was low [Lee, 1984].

Fairbanks and McGuire [1984] observed that contact aggression between females is rare; instead, dominance relationships seem to be maintained through preferential associations and threats. Daughters generally assume a rank just below their mothers, and at least in captivity, the ranking remains relatively stable over time [Fairbanks and McGuire, 1984]. In approximately 8 years, only three females of this captive group were treated for bite wounds. However, Else [1981] suggested that female fighting was a major problem in establishing vervet groups. Rowell [1971] stated that the relatively large canines of female guenons are effective weapons. The most common injuries are punctures and slashes near the root of the tail and elbow area. Gartlan and Brain [1968] recorded similar injuries, but in these cases the seemingly minor injuries led to extreme shock and death. The authors felt these deaths could have been related to stress from overcrowding.

Intergroup aggression may involve chases, but generally not intense threats or actual contact fighting. Intergroup meetings are fairly common, because there is a good deal of overlap between group home ranges [Dunbar, 1974; Hall and Gartlan, 1965; Lancaster, 1971; Poirier, 1972].

In wild vervets, Cheney et al [1981] have found that deaths due to illness were concentrated among low-ranking females, while high-ranking animals were more likely to be killed by predators. In the 3½-year study, ten of the

11 animals killed by predators were female. Severe competition for drinking water can be an especially important factor during the dry season. Wrangham [1981] reported the deaths of half of the adult females and two of the five infants, as well as two to four of the adult males in one group of vervets, apparently due to the lack of water. During the dry season, males invaded neighboring territories and one animal spent 10.8% of its activity budget engaged in activities related to drinking. In this study one male vervet was found dead with a broken neck, and other wounds on other dead males made it likely that they were killed by members of other groups.

Else [1981] has documented a 75% mortality rate among quarantined vervets who had just been taken from the wild. These animals developed severe diarrhea, became depressed, comatosed, and finally died. However, by delaying the usual quarantine manipulations such as TB testing, worming, and blood collection for 1 to 2 months, the mortality rate was dramatically decreased to 5–10%.

Else [1981] also reported eight infant deaths in 2 years. Two of these deaths were due to trauma, two to starvation, one to pneumonia, and two were of unknown cause. Young vervets are at the highest risk of death when they are under 5 months of age [Kushner et al, 1982]. Else stated that the high mortality rates of vervet monkeys can be overcome.

Cercopithecus mitis

In a field study, Rudran [1978] recorded only 99 aggressive interactions in 1,100 hours of observation. Adult males displayed the highest levels of aggression, and Rudran stated that this may be important in causing the emigrations of young males from their natal troops. Rudran also noted that group members fought most frequently and violently over food. Food may also have been an important influence on the incidence of intergroup conflicts. However, since the home ranges of blue monkeys did not overlap much, there was relatively little intergroup aggression. Intergroup conflicts that were observed involved adult, subadult, and juvenile monkeys of both sexes. Blue monkeys interacted with mangabeys, redtail monkeys, red colobus, and black and white colobus. Eighty-eight percent of these interspecific interactions were aggressive in nature, and subadult and juvenile blue monkeys were involved in these interactions more frequently than would be expected by chance [Rudran, 1978]. Chimpanzees are known to prey on blue monkeys and the blue monkeys appeared to avoid interactions with chimpanzees. Rudran reported the death of one infant that was abandoned by its mother after she was attacked by members of her group. Another death may also have been caused by intragroup aggression.

Cercopithecus mitis albogularis

Sykes' monkeys are similar to vervets in that the adult males are not necessarily the dominant animals in their groups [Rowell, 1971]. Aggression often occurs as a "chain reaction," and there is no control role played by dominant animals [Rowell, 1971]. Youngest animals are threatened the most by other group members. Rowell [1971] made the important observation that the lack of appeasement behaviors in the repertoire of Sykes' leads to a morbidity and perhaps even mortality problem in captivity. When aggressive interactions begin, they often escalate rather than being interrupted by an animal fleeing as might be the case in a wild environment. In the wild they probably run into dense vegetation [Rowell, 1971], and this may be an important reason to recreate areas of dense vegetation in captive environments for Sykes' monkeys. Mortality caused by intragroup aggression is apparently more common than in baboons, and Rowell [1971] reported five or six "murders" within the Sykes' colony. Such high levels of aggression may also affect reproductive behaviors. Rowell [1971] noted that two females that had the most stressful levels of social status did not breed, and were eventually killed by their group.

Erythrocebus patas

Patas monkeys are relatively less aggressive than many other species of African monkeys such as macaques and baboons [Hall, 1966 cited in Preston et al, 1970]. For example, in 627 hours of observation, Hall [1965] recorded only 49 agonistic interactions; only two of those involved biting. In captivity, higher levels of aggression are reported [Kaplan and Zucker, 1980]. Adult females are the major instigators and receivers of aggression. The harem male is involved in very little fighting and does not injure other group members [Hall et al, 1965; Kaplan and Zucker, 1980]. Similarly, very little aggression has been observed between all-male groups of patas [Gartlan, 1974]. Coalitions of females will intimidate the harem male if he threatens a young group member or is too close to a frightened infant [Fedigan, 1982]. Males and females may show different styles of aggression. Nagel and Kummer [1974] observed that a captive patas male threatened extensively before he attacked, whereas females sometimes would slowly approach another animal from behind and bite the animal.

The single most important cause of aggressive encounters among captive patas is the crying of an infant. This stimulus leads to the mother immediately attacking the perceived perpetrator of the infant's mistreatment [Nagel and Kummer, 1974]. A maturing male receives increasing levels of aggression from a mature male, and reintroductions of animals temporarily removed from the group may also be triggers to aggression [Hall and Mayer, 1967]. The distribution of food also seems to be an important factor in aggression. Kaplan and Zucker (1980) observed that the majority of fights occurred at the island's food provision areas. Hall [1965] discovered that where the natural food supply was restricted to small areas, aggressive conflicts increased. Patas express a clear tendency to feed alone [Kaplan and Zucker, 1980], and in captivity a lack of sufficient space to allow dispersed feeding may cause an increase in aggression. Like Sykes' monkeys, patas do not possess extensive repertories of "appeasement" signals (such as lip smacking or cowering), but instead seem to rely on physical avoidance to escape aggression [Kaplan and Zucker, 1980]. This inability to disperse widely may be a cause of higher levels of aggression in captivity than in wild settings, so it may be particularly important to avoid crowding in captive patas monkeys.

Miopithecus talapoin

Aggression may play an important role in the seasonally changing social composition of talapoins (see Reproduction section). Wolfheim [1977a] reported that spatial constraints in captivity may have led to the deaths of males in that study. These males were killed by older, female group members. In at least one case, all females of a group ranked higher than all of the males [Dixson et al, cited in Rowell and Dixson, 1975]. Female talapoins engage in more aggressive interactions than do males, and the females may "gang up" on males during agonistic encounters. Males show higher levels of aggression toward other females when one female is at midcycle [Scruton and Herbert, 1970].

GROUP FORMATION

Cercopithecus aethiops

Early attempts at establishing captive vervet groups were unsuccessful. High levels of fighting disrupted attempts to introduce adult animals into infant and juvenile groups [Gartlan and Brain, 1968]. According to Mallinson [1971], it appeared to be impossible to integrate adult vervets in captivity, and Rowell [1971] reported some deaths from fighting as a captive group was being established. But Gartlan and Brain [1968] also described the successful formation of a vervet troop at the Johannesburg Zoo in 1964. Fifty-seven animals were put into a large, circular aviary at the same time. Although there was initial aggression, the animals were able to avoid close contact because of the size of the area. Even with 30 males in the group, extremely high levels of aggression were never recorded.

Else [1981] described in some detail the management methods used at the Institute for Primate Research near Nairobi, Kenya. Incoming animals are quarantined in individual cages for 3 months (see Mortality Section for deaths during quarantine). Following the quarantine, vervets are placed in harem groups with one male and five to ten females. Any female that is seriously injured or excessively harassed is put into a different harem group. The male infants from the groups are removed at 3 months of age and are put into weanling groups, but the infant females are left to increase the size of the harems. The adult males are exchanged into other groups when the female progeny becomes 2 years old. These exchanges did not lead to any aggression problems.

Group formation was characterized as "difficult" because of female fighting, and each group was reduced by two to four females over the first year of group formation. Two of the females that died (due to pneumonia and colitis) had been the victims of fairly continual harassment by cagemates. It was at least 1 year after group formation before females stopped fighting. In retrospect, Else suggested placing animals from the same wild troops together if possible, or establishing weanling groups.

In one field study, 67% of recorded injuries were to males that were new arrivals to a group. Although they were the targets of much aggression, they were not chased away. In this case, juvenile and female group members were the first to engage in nonaggressive contact with newly arriving males [Henzi and Lucas, 1980].

One experimental analysis of the effects of crowding vervets has been published [McGuire et al, 1978]. In forming new groups of juvenile animals, the investigators were surprised to find that familiarity did not appear to effect their response to social density. In a second experiment, it was discovered that agonistic behavior did *not* increase with crowding, but affiliative behaviors did increase.

Miopithecus talapoin

Scruton and Herbert [1972] found that introducing stranger talapoins into groups of six or fewer animals always led to high frequencies of aggressive episodes directed to the newly introduced animal. Most of this aggression was initiated by group members that were of the same sex as the newcomer. There was no evidence that the estrous state of a female newcomer conferred any protection from aggression.

ENRICHMENT/FEEDING STRATEGIES

Cercopithecus aethiops

Vervets are opportunistic omnivores that consume leaves, fruit, flowers, insects, bird eggs, crustaceans, and lizards [Fedigan, 1982; Wrangham, 1981]. Although they eat a wide variety of food items, the extreme seasonality of their natural habitat causes long periods of dietary monotony. However, there is a rapid exploitation of new food sources once they become available [Kavanaugh, 1980]. Poirier [1972] noted that some vervet food sources require preparation before eating. For example, the shell surrounding the cockrass bean is removed before the jelly covering of the seed is eaten. The skins of mangos, clammy cherries, and breadfruits are peeled off with teeth and hands before the fruits are consumed. Vervets are known as ''messy'' or ''wasteful'' feeders, but one study has shown this feeding style serves a social function, as it ensures a distribution of food among all group members [Tollman and Simpson, 1975]. These authors also discovered that aggression within the group seemed higher when food was presented in one pile than when it was presented in four piles. This impression should be subjected to experimental analysis. Fairbanks and McGuire (1985) attempted to simulate natural conditions in their outdoor colony. The feeding routine consisted of continually available ''monkey chow'' and the opportunity to forage in grass. Occasional fruits and vegetables were also provided.

Cercopithecus mitis stuhlmanni

C. mitis feed on a wide variety of plant species and invertebrates. Rudran [1978] reported that one group of blue monkeys consumed at least 59 plant species and 101 specific food items. Invertebrates such as butterflies, grasshoppers, and ants constituted almost 20% of the total feeding observations.

Fairbanks [1984] reported 61 instances of vertebrate predation in a captive colony over a 6-year period. Approximately equal numbers of rodents and birds were killed, but birds were preferred for consumption. Most observed captures were made by juvenile males and females.

Erythrocebus patas

The patas diet generally consists of grasses, seeds, fruits, fungi, leaves, flowers, lizards, insects, and bird eggs [Fedigan, 1982; Gartlan, 1974; Hall et al, 1965]. Free-ranging patas disperse while feeding, and sufficient space to allow feeding alone should be made available to captive patas as well. There is indirect evidence for food competition, as Hall et al [1965] found that in the laboratory only subordinate young males distended their cheek pouches with food. Perhaps this could be avoided with ample opportunity for solitary feeding.

Koster [1985] reported that a single female patas monkey in West Africa ate 23 different species of leaves, fruit, flowers, saps, and seeds. She also ate animal material such as termites, ants, grasshoppers, lizards, and birds.

Olson [1965] considers patas like other anthropoid species, to be true "gum specialists," as the gum exudate of *Acacia* is a major dietary component. At least 20% of the adult feeding activity per month is spent gum feeding by wild patas in Kenya. This would be a profitable behavior to simulate in captive settings through the use of manipulanda containing gumlike consumable substances.

Miopithecus talapoin

Gauthier-Hion [1970] noted that agonistic encounters were quite common during feeding times in the wild talapoin group observed. Highly localized food sources were associated with maximum levels of aggression.

DIMENSIONS OF HABITAT

Our review of field and laboratory research has provided a framework for understanding the innate propensities of the *Cercopithecines*. We have highlighted those behavioral systems that are most important for managing and breeding many of these taxa in captive settings. It will also be useful to identify the key variables that are likely to control captive primate behavior. Since the subjects of this review have not been studied in this context, we must apply findings from studies of diverse primate taxa.

Spatial Variables

Because of their relatively greater intellectual powers, stimulus deprivation is regarded as a serious problem for captive apes [e.g., Maple, 1979]. However, monkey behavior patterns are also disrupted when the physical or social environment is deficient or aversive. For example, Glatson et al [1984] found that cotton-topped tamarins behaved inappropriately when they were housed within the public view. While on exhibit, these tamarins emitted fewer social behaviors, engaged in greater amounts of parental aggression, and played and mated to a lesser extent. Aggression within the group was correlated with the number of visitors present when the animals were on display. To remedy these problems, the authors suggested quieter cage situations with more emphasis on appropriate furniture and cover. Recommendations such as these would doubtless apply to similar situations where African monkeys were concerned. Our literature review indicates that some taxa would be more vulnerable to intrusive exhibition conditions than others.

When a monkey is exhibited in close proximity to zoo visitors, the situation may be so aversive that normal social behaviors are compromised. Under such living conditions, young animals are quite likely to develop stereotyped motor patterns and abnormal deprivation acts such as autoerotic behaviors and/or self-aggression. Such behaviors are unlikely if the animal is permitted some control over its distance.

The African monkeys that we have examined in this chapter are well adapted for living or at least spending some of their time in trees. It is advisable, therefore, to maintain these taxa in environments that permit movement in the vertical dimension. For African monkeys, volumetric space is a means to provide cover, privacy, and social distance. By providing a usable volume within a small enclosure, the deleterious effects of crowding can be effectively reduced. This is accomplished by providing tree branches, ropes, cargo nets, perimeter ladders, elevated cage furniture (e.g., benches or sleeping platforms), and other vertical opportunities.

Naturalistic Habitats

The environmental psychologist Robert Sommer [1972] used the terms "hard architecture" to label barren, oppressive, and restricted settings such as prisons, mental hospitals, and many zoos. According to Sommer, both animals and people are degraded by such living environments. As he stated in his essay: "Confinement under oppressive conditions results in adaptations that are rarely in the best interests of the confined individuals or society" (p. v).

The contrast between "barren" and "enriched" environments has also been discussed by Wemelsfelder [1984], who wrote: "The environment provides adequate stimuli to fulfill an animal's needs, and elicits appropriate, species-specific behaviour in an animal. In a diverse, rich environment, an animal has to search for these adequate stimuli. . . . In a monotonous, "barren" environment, the animal is not only prohibited from searching for adequate stimuli most of the time, but hardly any adequate stimuli are present" (p. 121).

Naturalizing elements such as browse have been evaluated by Chamove et al [1982], whose team discovered that the addition of deep straw wood chip litter reduced aggression and increased the activity of captive monkeys. Deep straw conceals small food items such as raisons, seeds, or popcorn and encourages a kind of foraging activity. This form of cage enrichment may be employed to combat boredom and the development of deprivation acts. The positive effects of such enrichment procedures have been more completely reviewed elsewhere [Maple, 1979; Maple and Finlay, 1986].

An important aspect of the naturalized environment is the degree to which the environment can be manipulated. For example, Westergaard and Fragaszy [1985] determined that capuchin monkeys dramatically increased their object-directed activities after movable materials were introduced. These investigators concluded that the opportunity to express species-typical manipulative tendencies was conducive to the monkeys' physical health and psychological well-being.

Monkeys that reside in zoos should be able to express the full range of their behavioral potential. This potential has been characterized by others [e.g. Rollin, 1981] as the basic nature, or "telos," of the animal. Appropriate naturalistic environments encourage the expressions of natural behavior patterns. When the amount of space is optimal, the living environment complex and variable, and the social grouping appropriate, the behavior of the captive animal will closely resemble its wild counterpart. Zoos cannot educate properly if zoo animals do not emit a repertoire of species-typical behaviors.

Another effective method for enriching captive environments is to provide for contingencies of reinforcement. With this technology it is possible to induce activity (even natural behavior such as prey-acquisition) with a scheduled and contingent reward system.

Toward "Actualized" Monkeys

"Self-actualization" refers to the process of reaching one's full potential. The full potential of monkeys and apes can also be "actualized." However, captive monkeys depend on the creativity of human designers and managers to reach their full primate potential. By fully understanding nonhuman primates within the context of their natural habitat and history, it will be possible to manage them in optimal settings. By so doing we will produce colonies and collections of animals that will mate, parent, and express their behavioral repertoire in the appropriate fashion of their kind.

Successful simulations of natural primate habitat have been achieved in a number of captive settings worldwide. Primate reproduction cannot fail to improve in such enlightened and enriching surroundings.

TAXONOMIC SUMMARIES

Cercopithecus aethiops

In this taxon it appears that a strong dominance hierarchy increases group cohesion. These monkeys reside in multimale/multifemale groups in the wild. Members of the group occupy trees and rock formations to engage in vigilant behavior. Males become troublesome in their natal group between the ages of 4 and 6. It may also be advisable to remove alpha males after a 3–5 year reign. Wild individuals do not begin to cycle regularly in captivity for 6–12 months. This taxon reproduces well in captivity, at a rate that is faster than macaques. They apparently reproduce under diverse conditions. It appears that *C. aethiops* thrives best in complex social groupings. However, American zoos have tended to manage this taxon in small family or monogamous units (Table I). Because of their very large canine teeth, female fighting can be just as serious as male encounters.

Cercopithecus mitis spp.

In the Sykes' subspecies several years are apparently needed for a wild animal to fully adjust to a captive setting. These taxa breed later than macaques or *C. aethiops*. However, *C. mitis* spp. appear to engage in breeding well into advanced age, as compared to macaques and other Old World monkeys. In the wild, Sykes' monkeys settle conflicts by escaping into the thick vegetation. Because male Sykes' monkeys are reluctant to control aggression within the group, agonistic encounters tend to escalate in a captive setting. Accordingly, it is advisable to provide enclosure furniture, cover, and adequate space for this taxon in captivity.

Erythrocebus patas

Living primarily in harem groupings, wild males tend to be peripheral to the group. In fact, this taxon prefers to maintain an unusually large home range, and they require a large intergroup distance from conspecifics. In captivity, two groups of patas should be separated by another species in a buffer environment. Patas are quite active and reside in very large groupings. The male may need a separate sleeping cage at night. In this taxon, kidnapping can be a problem; adequate cover would help reduce this tendency. Experts advise that new mothers should nevertheless remain in the group and not be isolated. The young of this taxon are precocious. Although they reside in large groups in the wild, care should be taken to avoid high social density in captivity. To reduce competition, dispersed feeding should be practiced.

Miopithecus talapoin

This very fertile primate tends to occupy single sex subgroups in the wild. The sexes should be brought together only for mating. After pregnancy the females should be permitted adequate social distance from cagemates.

Cercopithecus neglectus

The social organization of this taxon is still unclear. While recent evidence from Kenya indicates a tendency to monogamy, this would represent an unusual propensity for *Cercopithecus* and indeed the entire Old World monkey family. It is certainly possible that the monogamous tendency in Kenya is due to habitat destruction and the resultant dispersion of individuals. It is abundantly clear from the data in Table III that American zoo professionals do not universally regard this taxon as monogamous. *C. neglectus* has successfully reproduced in both conditions.

ACKNOWLEDGMENTS

The authors gratefully acknowledge research support from the following sources: 1) NIH grant RR-00165 from the Division of Research Resources to the Yerkes Primate Research Center of Emory University; 2) NIH contracts N01-RS-7142 and N01-RS-3-2116 to the University of Texas System Cancer Center; and 3) small grants from Wildlife Preservation Trust International, Atlanta Zoological Society, and Atlanta/Fulton County Zoo, Inc. to the second author. For assistance during the preparation of this manuscript the authors thank Wanda Abbott, Lorraine Perkins, and Ted W. Finlay.

REFERENCES

Aldrich-Blake FPG (1970): Problems of social structure in forest monkeys. In Crook JH (ed): "Social Behaviour in Birds and Mammals." London: Academic Press, pp. 79–101.

Bramblett CA, Pejaver LD, Drickman DJ (1975): Reproduction in captive vervet and Sykes' monkeys. J Mammal 56:940–946.

Bramblett CA, Bramblett SS (1982): Seasonality of births among captive vervet monkeys. Am J Phys Anthropol 57:171.

Brandt EM, Mitchell G (1971): Parturition in primates: Behavior related to birth. In Rosenblum LA (ed): "Primate Behavior," Vol. 2. New York: Academic Press, pp. 177–223.

Brennan EJ (1985): DeBrazza's monkeys in Kenya: Census, distribution, and conservation. Am J Primatol 8:269–277.

Brennan EJ (in press): Demographics of captive DeBrazza's guenons. Zoo Bio.

Brennan EJ, Else JG (1984): Can Kenya's DeBrazza monkeys survive? Swara 73:12–14.

Butynski TM (1982): Harem male replacement and infanticide in the blue monkey (*Cercopithecus mitis stuhlmanni*) in the Kibale Forest, Uganda. Am J Primatol 3:1–22.

Chamove AL, Anderson JR, Morgan-Jones SC, Jones SP (1982): Deep woodchip litter: Hygiene, feeding, and behavioral enhancement in eight primate species. Int J Study Anim Prob 3, 4:308–319.

Cheney DL, Lee PC, Seyfarth RM (1981): Behavioral correlates of non-random mortality among free-ranging female vervet monkeys. Behav Ecol Sociobiol 9:153–161.

Chism J (1986): Development and mother-infant relations among captive patas monkeys. Int J Primatol 7:49–81.

Chism J, Rowell TE, Richards SM (1978): Daytime births in captive patas monkeys. Primates 19:765–767.

Cords M, Rowell TE (1986): Group fission in blue monkeys of the Kakamega Forest, Kenya. Folia Primatol 46:70–82.

Dunbar RIM (1974): Observations on the ecology and social organization of the green monkey, *C. sabaeus,* in Senegal. Primates 15:341–350.

Ehardt-Seward C, Bramblett CA (1980): The structure of social space among a captive group of vervet monkeys. Folia Primatol 34:214–238.

Else J (1981): Captive breeding of vervet monkeys (*C. aethiops*) in harems. Reproduction 8:1–9.

Else JG, Eley RM, Suleman MA, Lequin RM (1985): Reproductive biology of Sykes' and blue monkeys (*Cercopithecus mitis*). Am J Primatol 9:189–196.

Else JG, Eley RM, Worthman WC, Lequin RM (1986): Reproduction in the vervet monkey (*Cercopithecus aethiops*): II. Annual menstrual patterns and seasonality. Am J Primatol 11:333–342.

Fairbanks LA (1984): Predation by vervet monkeys in an outdoor enclosure: The effect of age, rank, and kinship on prey capture and consumption. Int J Primatol 5:263–273.

Fairbanks LA, McGuire MT (1984): Determinants of fecundity and reproductive success in captive vervet monkeys. Am J Primatol 7:27–38.

Fedigan LM (1982): "Primate Paradigms: Sex Roles and Social Bonds." Montreal: Eden Press.

Gartlan JS (1969): Sexual and maternal behaviour of the vervet monkey, *Cercopithecus aethiops.* J Reprod Fertil, Suppl. 6:137–150.

Gartlan JS (1974): Adaptive aspects of social structure in *Erythrocebus patas.* In "Symposium Vth Congress of the International Primatological Society," pp. 161–171.

Gartlan JS, Brain CK (1968): Ecology and social variability in *C. aethiops* and *C. mitis.* In Jay P (ed): "Primates: Studies in Adaptation and Variability." San Francisco: Holt, Rinehart and Winston, pp. 253–292.

Gautier-Hion A: L'organization sociale d'une bande de talapoin dans le nord-est du Gabon. *Folia Primatologica* 12:116–141.

Glatson AR, Geilvoet-Soeteman E, Hora-Pacek E, van Hooff JARAM (1984): The influence of the zoo environment on social behavior of groups of cotton-topped tamarins, *Saguinus oedipus oedipus.* Zoo Biol 3:241–254.

Goswell MJ, Gartlan JS (1965): Pregnancy, birth and early infant behaviour in the captive patas monkey, *Erythrocebus patas.* Folia Primatol 3:189–200.

Hall KRL, Gartlan JS (1965): Ecology and behaviour of the vervet monkey, *Cercopithecus aethiops,* Lolui Island, Lake Victoria. Proc Zool Soc Lond 145:37–56.

Hall KRL, Boelkins RC, Goswell MJ (1965): Behaviour of patas monkeys, *Erythrocebus patas,* in captivity, with notes on the natural habitat. Folia Primatol 3:22–49.

Hall KRL, Mayer B (1967): Social interactions in a group of captive patas (*Erythrocebus patas*). Folia Primatol 5:213–236.

Harding RSO, Olson DK (1986): Patterns of mating among male patas monkeys (*Erythrocebus patas*) in Kenya. Am J Primatol 11:343–358.

Hartwell KM (1978): Reproductive cycles of patas monkeys, (*Erythrocebus patas*) with special reference to breeding seasons. Dissertation Abstr Int B 38, 8:3585–3586.

Henzi SP, Lucas JW (1980): Observations on the inter-troop movement of adult vervet monkeys (*C. aethiops*). Folia Primatol 33:220–235.

Hess DL, Hendrickx AG, Stabenfeldt GH (1979): Reproductive and hormonal patterns in the African green monkey (*Cercopithecus aethiops*). J Med Primatol 8:273–281.

Hill WCO (1966): Laboratory breeding, behavioural development and relations of the talapoin (*Miopithecus talapoin*). Mammalia 30:353–374.

Johnson PT, Valerio DA, Thompson GE (1973): Breeding the African green monkey, *Cercopithecus aethiops*, in a laboratory environment. Lab Anim Sci 23:355–359.

Johnson C, Koerner C, Estrin M, Duoos D (1980): Allopaternal care and kinship in captive social groups of vervet monkeys (*Cercopithecus aethiops sabaeus*). Primates 21:406–415.

Kaplan JR, Zucker E (1980): Social organization in a group of free-ranging patas monkeys. Folia Primatol 34:196–213.

Kaplan JR, Anthony M, Wood L (1981): Domestic breeding of patas monkeys (*Erythrocebus patas*). Lab Anim Sci 31:409–412.

Kavanaugh M (1980): Selective pressures and the socio-ecology of a West African Cercopithecine monkey. Trop Ecol Dev 21:377–382, 1980.

Keddy AC (1986): Female mate choice in vervet monkeys (*Cercopithecus aethiops sabaeus*). Am J Primatol 10:125–134.

Kingdon J (1971): "East African Mammals," Vol 2. New York. Academic Press.

Koster SH (1985): Food habits of a free ranging patas monkey (*Eurythrocebus patas*) in "W" National Park, Niger. Mammalia 49:589–591.

Kushner H, Kraft-Schreyer N, Angelakos ET, Wudarski EM (1982): Analysis of reproductive data in a breeding colony of African green monkeys. J Med Primatol 11:77–84.

Lancaster JB (1971): Play-mothering: The relations between juvenile females and young infants among free-ranging vervet monkeys (*Cercopithecus aethiops*). Folia Primatol 15:161–182.

Lee PC (1984): Ecological constraints on the social development of vervet monkeys. Behaviour 91:245–262.

Loy J, Head M, Loy K (1978): Reproductive cycles of captive patas monkeys. Lab Primate Newsletter 17:9–12.

Mallinson JJC (1971): Observations on the reproduction and development of vervet monkey with special reference to intersubspecific hybridization. Mammalia 35:598–609.

Maple TL (1979): Great apes in captivity: The good,the bad, and the ugly. In Erwin J, Maple TL, Mitchell G (eds): "Captivity and Behavior." New York: Van Nostrand Reinhold Co., pp. 239–272.

Maple TL, Finlay TW (1986): Evaluating the environments of captive nonhuman primates. In Benirschke K (ed): "Primates: The Road to Self-Sustaining Populations." New York: Springer Verlag, pp 479–488.

McGuire MT, Cole SR, Crookshank C (1978): Effects of social and spatial density changes in *Cercopithecus aethiops sabaeus*. Primates 19:615–631.

Nagel U, Kummer H (1974): Variation in cercopthecoid aggressive behavior. In: "Primate Aggression, Territoriality and Xenophobia." New York: Academic Press, pp. 159–184.

Olson DK (1985): Gum exudates as a dietary staple of Patas monkeys. Am J Phys Anthro 66:210.

Poirier FE (1972): The St. Kitts green monkey (*Cercopithecus aethiops sabaeus*): Ecology, population dynamics, and selected behavioral traits. Folia Primatol 17:20–55.

Preston DG, Baker RP, Seay B (1970): Mother-infant separation in the patas monkey. Dev Psychol 3:298–306.

Richman D (1963): Husbandry for successful breeding of patas monkeys (*Erythrocebus patas*). J Anim Tech Assoc 14:31–35.

Rollin BE (1981): "Animal Rights and Human Morality." Buffalo, New York: Prometheus Books.

Rowell TE (1970): Reproductive cycles of two *Cercopithecus* monkeys. J Reprod Fertil 22:321–338.

Rowell TE (1971): Organization of caged groups of *Cercopithecus* monkeys. Anim Behav 19:625–645.

Rowell TE (1977): Reproductive cycles of the talapoin monkey (*Miopithecus talapoin*). Folia Primatol 28:188–202.

Rowell TE, Dixson AF (1975): Changes in social organization during the breeding season of wild talapoin monkeys. J Reprod Fertil 43:419–434.

Rowell TE, Hartwell KM (1978): The interaction of behavior and reproductive cycles in patas monkeys. Behav Biol 24:141–167.

Rowell TE, Richards SM (1979): Reproductive strategies of some African monkeys. J Mammal 60:58–69.

Rudran R (1978): Socioecology of the blue monkeys (*C. mitis stuhlmanni*) of the Kibale Forest, Uganda. Smithsonian Contrib Zool 249:1–88.

Schapiro S (1984): Reproductive synchronization in *Cercopithecus aethiops* and *C. mitis*. Paper presented at the meeting of the American Society of Primatologists, Humboldt, California, June 1984.

Scruton DM, Herbert J (1970): The menstrual cycle and its effect on behaviour in the talapoin monkey (*Miopithecus talapoin*). J Zool 162:419–436.

Scruton DM, Herbert J (1972): The reaction of groups of captive talapoin monkeys to the introduction of male and female strangers of the same species. Anim Behav 20:463–473.

Sommer R (1972): What do we learn at the zoo? Nat Hist, LXXXI:26–29, 84–85.

Stevenson MF (1973): Observations of maternal behaviour and infant development in the DeBrazza monkey in captivity. Int Zoo Yearbook 13:179–184.

Struhsaker TT (1967): Social structure among vervet monkeys (*C. aethiops*). Behaviour 29:83–121.

Tollman SG, Simpson JRE (1975): An examination of the apparently "wasteful" feeding behaviour in *Cercopithecus aethiops:* A new perspective? J Behav Sci 2:99–114.

Tsingalia HM, Rowell TE (1984): The behaviour of adult male blue monkeys. (Zeitschrift Fur Tierpsychologie) 64:253–268.

Valerio DA, Johnson PT, Leverage WE, Thompson GE (1972): Laboratory breeding and husbandry of macaques, African green monkeys and greater bushbabies. In Goldsmith EI, Moor-Jankowski J (eds): "Medical Primatology 1972," Part I. Basel: S. Karger, pp 156–165.

Wemelsfelder F (1984): Animal boredom: Is a scientific study of the subjective experiences of animals possible? In Fox MW, Mickley LD (eds): "Advances in Animal Welfare Science." Washington, D.C.: Humane Society of the United States, pp 115–154.

Westergaard GC, Fragaszy DM (1985): Effects of manipulable objects on the activity of captive capuchin monkeys (*Cebus apella*). Zoo Biol 4:317–328.

Wolfheim JH (1977a): A quantitative analysis of the organization of a group of captive talapoin monkeys (*Miopithecus talapoin*). Folia Primatol 27:1–27.

Wolfheim JH (1977b): Sex differences in behavior in a group of captive juvenile talapoin monkeys (*Miopithecus talapoin*). Behaviour 63:110–128.

Wrangham RW (1981): Drinking competition in vervet monkeys. Anim Behav 29:904–910.

Zucker EL, Kaplan JR (1981): Allomaternal behavior on a group of free-ranging patas monkeys. Am J Primatol 1:57–64.

Index